Spatial policy problems of the
British economy

Spatial policy problems of the British economy

edited by **MICHAEL CHISHOLM**
Reader in Geography, University of Bristol

and **GERALD MANNERS**
Reader in Geography, University College London

CAMBRIDGE at the University Press 1971

Published by the Syndics of the Cambridge University Press
Bentley House, 200 Euston Road, London NW1 2DB
American Branch: 32 East 57th Street, New York, N.Y.10022

© Cambridge University Press 1971

Library of Congress Catalogue Card Number: 70–160090

ISBN: 0 521 08235 8

Printed in Great Britain
at the University Printing House, Cambridge
(Brooke Crutchley, University Printer)

Contents

Tables

Tables

Figures*

Figures

* *All the figures for this volume were drawn by Miss Margaret E. Thomas in the Cartographic Unit of the Department of Geography, University College London.*

Preface

In preparing this collection of essays, we have tried to ensure that a coherent theme runs through all the contributions, namely, the importance of the spatial dimension in the formulation of public policy. Though the volume deals explicitly only with Britain, we are confident that readers will recognise the general nature of the issues that are raised, which have not in the past received their due recognition.

Apart from the common interest of the authors in this particular field of enquiry, there is another bond that we should like to record. All the contributors pursued their undergraduate studies in geography at St Catharine's College, Cambridge, under the guidance of Mr A. A. L. Caesar following his return to the College twenty years ago in 1951. It was under his guidance that all the contributors were first introduced to spatial policy problems of the British economy.

M.C.
G.M.

1. Geographical space: a new dimension of public concern and policy

MICHAEL CHISHOLM AND GERALD MANNERS

The last decade has witnessed a growing public interest in the geography of economic and social change. The emotions aroused by the prospective impact and cost of the London motorway box, the vigour with which Scottish and Welsh interests canvassed for an aluminium smelter and the conflicting interests exposed by the Roskill Commission's search for a site suitable as an international airport are common currency amongst informed opinion today. The issues may at first sight appear diverse. Yet underlying the continuing debate on what should be done to counter the persistently high levels of unemployment in the less prosperous parts of the country and the much publicised 'drift to the south', and surrounding the many arguments over shifts in local or sub-regional patterns of land use together with the frequently associated issues of environmental quality, there is a single unifying theme. This is the uncertainties, for individuals and corporate groups, inherent in making decisions concerning the spatial allocation of resources and activities. This new focus of interest and debate, which has given a renewed importance to the study and interpretation of geographical space, stems from the convergence of four developments in particular.

Social versus spatial inequalities

Great advances have been made in Britain during the present century in the attempt to equalise opportunities and conditions between different socio-economic classes. Poverty and inequality still exist, to a degree that varies according to the definitions employed. But the undoubted achievements of the welfare state in demolishing the principal bastions of inequality have exposed more vividly than ever before other causes for equalitarian public concern, amongst which are several characterised by their spatial as much as by their social nature.

Inter-regionally, there has emerged a continuing imbalance between the economic and social opportunities available in the southern and midland zones of Britain on the one hand and the less prosperous counties of northern England, Scotland and Wales on the other. For thirty-five years, attempts have been made by governments to check the out-migration, to provide new sources of employment and to improve the long-term economic prospects of the older, coalfield-based regions of Britain. In these areas, structural economic changes have occurred – and continue – on a scale and at a rate which would inevitably result in widespread social distress were they to be left solely to the mechanisms of the market. In response, trading estates and advanced factories have been used to attract private manufacturing enterprise; subsidies have been offered to support the old and to encourage the new industries located there; subventions have been given to employers through such measures as the Regional Employment Premium (Department of Economic Affairs, 1967; First Secretary of State, 1967); and the transfer of Civil Service jobs has yet further diversified their economic structure. These measures have all sought as an immediate objective to reduce the level of unemployment – the most blatant symptom of the less prosperous regions' ills – and to provide a better economic future for the people living there. The economic problems of these areas have been reflected in a (public and private) failure adequately to renew their built environment. The result, as the Hunt Committee (Secretary of State for Economic Affairs, 1969) stressed, has been the creation of a parallel and distinctive geography of amenity, taking amenity in the widest sense. In education, health, housing and cultural facilities, the less prosperous parts of Britain enter the nineteen-seventies with a relative disability which is a reflection of, and is mirrored in, the quality of life and the nature of individual opportunity there. The more prosperous south and midlands of Britain are certainly not without their problem localities. Nevertheless, their comparative advantages, attractiveness and success are all too apparent in terms of such indices as the range of employment opportunities and income growth, the large proportion of growth industries, high activity rates and net in-migration, and the superior range of communication facilities and generally more attractive urban environment to be found there. These characteristics are firmly imprinted upon the public mind.

Intra-regionally, also, differences in prosperity have come to be accentuated through time. Within the less prosperous regions, cities such as Newcastle and Cardiff, Liverpool, Leeds and Glasgow have been able to embark upon major schemes of urban redevelopment which, in time, will considerably enhance their reputations as provincial centres and improve the quality of their citizens' lives. Agglomeration economies, together with deliberate government policies, have ensured that the range and stability of their economic opportunities have increased. This in turn has its social implications. In contrast, although they have not been without small-scale private and somewhat larger public investment, it is likely that the villages and small towns of (for example) the Durham and Glamorgan coalfields, of the Lancashire and Yorkshire textile communities and of many parts of the central valley of Scotland will have to accept a gradual disappearance of their local economic base and to live with their Victorian urban legacy and dwindling social assets for many years to come. Sometimes, the realities of the situation are publicly faced. Durham County Council has bravely listed its category 'D' villages, which are due to be erased from the landscape. Elsewhere, such boldness has been lacking.

Simultaneously, intra-regional contrasts of wealth and opportunity have been growing within the more prosperous midlands and the south. In addition to the long-standing rural–urban differential in both economic opportunity and life style, contrasts between the housing standards, educational facilities and social provisions in certain central parts of the conurbations and central cities on the one hand and their suburbs on the other have become increasingly vivid. As Pahl points out in Chapter 5, the typical low-paid worker in the nineteen-seventies is much more likely to be employed in an urban service job than in manufacturing or primary industry. The situation has been complicated and given an emotional dimension by the arrival of coloured immigrants in the central areas of cities and the emergence of racially segregated quarters. These spatial contrasts are prospectively so threatening to the future social stability of the city that they have generated a public response in the form of accelerated urban renewal programmes, and additional funds for 'educationally deprived' areas within large urban areas.

The growing public awareness of areas of relative deprivation – an appreciation that the dilemma of the 'two nations' has shifted from being a class to being to an important degree a spatial problem, and a recognition that zones of urban blight are as much a technological and urban planning dilemma as they are a social and racial problem – has been accompanied by a much less acrimonious debate than the one that preceded the major welfare legislation of Victorian and Edwardian times. In the last resort, pleas for greater economic and social equality as between *places* are somewhat less convincing than appeals for greater economic and social equality as between *men*. Nevertheless, it has been remarkably widely agreed that some solutions to the problems must be sought through public action – although the precise nature of the programmes to be adopted has naturally been the subject of considerable and sometimes heated debate. The uncertainties facing the policy-makers are clear. Should the emphasis of public policy fall upon getting more jobs into the less prosperous parts of the country, or upon encouraging intra- and inter-regional population mobility? Should the problems of the inner cities be solved primarily through programmes of rehabilitation *in situ*, or should a greater emphasis be placed upon the encouragement of the selective movement of relatively deprived communities into strategically placed reception localities in and beyond the suburbs? To the extent that the arguments rest upon such complex matters as the mobility of industry, the nature and magnitude of external economies, the economics of urban renewal and the sociology of population mobility, they have been removed from the arenas of the most acrimonious party political argument. Increasingly, cost effectiveness studies have been pursued to offer improved guidelines for public policy; ways of checking the mounting expense of regional development policies is a case in point. Such studies do not remove the problems from the public eye or ensure equity. Nor do they remove completely the need for political judgement in establishing the priorities for government action, in formulating the assumptions underlying any analysis, or in weighing the more theoretical solutions against the range of practical possibilities. However, they do place a new importance upon attempts to improve the quality of spatial (economic and social) analysis.

The widened role of government

The long-term historical tendency has been for central and local government together to spend an increasing share of the Gross National Product. Before World War I, the proportion was under 15 per cent; it had more than doubled to 30 per cent by 1938, had risen to 40 per cent by the mid-nineteen-sixties (Chisholm, 1970, 196) and in the last full year of the recent Labour Administration to about half the national income. The Conservative government elected in 1970 is pledged to reduce public spending relative to the growing wealth of the country, but it will clearly have to take quite drastic action even to hold the proportion constant, let alone reduce it. However successful government may be in reversing the long-term trend, it is nevertheless generally accepted in political circles that public decisions and expenditure will continue to play a major role in the evolution of economic and social affairs in the country. In turn this means that the government will continue to possess a very considerable leverage upon the location of economic activities.

At the core of public intervention in the evolution of the country's spatial (economic and social) systems is a closely inter-related group of decisions affecting transport facilities and land use which it is impossible for a modern government to avoid. The public ownership of the railways apart, it is clearly the responsibility of government to supervise the evolution of both inter-regional and intra-regional transport facilities in general and roads in particular. Roads afford not only the primary means of spatial interaction, but they also condition the framework within which economic and social life evolves. Whilst it is perfectly clear that in the past decisions on road investments have only occasionally been seen in this light, and that only in the late nineteen-sixties did the Ministry of Transport (now part of the Department of the Environment) begin thinking of the country's future road network in system terms (Ministry of Transport, 1969), it is being increasingly recognised that the inter-city motorway programme will be one of the more influential forces shaping the economic geography of the country in the nineteen-seventies and nineteen-eighties. And the approach prospectively to be adopted towards the geographical priorities of urban motorway construction will rank among the more important determinants of city form and urban life style of our children.

The implications of major transport improvements for the pattern of sub-regional development have only in recent years been more fully appreciated. The construction of a motorway between two towns, for example, results in much more than simply the expedition of traffic between them; it has wide-ranging implications for their economic and social life – altering the meaning of distance and encouraging an upward movement in the levels and intensity of interaction between them – as well as modifying the relative economic and social prospects of the communities which it by-passes without affording easy access. Although the full impact of major new estuarine road crossings, such as those recently built to span the Forth and the Severn, is even now not fully quantified, there can be little doubt concerning the revolution which is engendered in the spatial relationships of the communities on both sides (Manners, 1966). The economic development implications of such investments must therefore be recognised and taken into account in programmes of regional and sub-regional growth (Central Unit for Environmental Planning, 1969).

Other elements in the country's communications system – airports, seaports, railways and telecommunication facilities – have in their several ways equally far-reaching spatial implications. There can be little doubt that, even though the mechanisms of public responsibility can vary from time to time and from place to place, the provision of such facilities is ultimately the inescapable responsibility of government. The implications of such infrastructural investments for the geography of the country's economic life – the fact that they not only help to mould the future pattern of space relationships, but that they must serve the existing one efficiently – demand that the government's role in the provision and supervision of transport is associated with a further set of responsibilities relating to the use of land.

It was as long ago as 1903 that the first legislation was passed to assert a public interest in the evolution of urban land uses. Since then, public involvement in the allocation of land to new uses has increased considerably, both in terms of its more widespread application, and in the geographical scale to which it has applied. The county and county borough plans prepared under the 1947 Town and Country Planning Act; the designation of new towns, expanded towns and then more recently new cities with a planned population of 250,000; the publication of sub-regional and regional studies and strategies in the nineteen-sixties; and the development of metropolitan transport and land-use expertise and plans have all expressed the need for enlarging the spatial scale of the public role in the allocation of land uses. Indeed, the prospective magnitude of population increase and urban expansion has been judged to warrant both large injections of central government funds and initiatives into regions blessed with outstanding assets for accelerated growth and population absorption, and the careful supervision of the considerable changes which will be necessary in the use of land there. Humberside, Tayside and Severnside were each examined in this context by the Central Unit for Environmental Planning or the Scottish Development Department between 1966 and 1971.

The precise means of public intervention in the processes of land-use change are, of course, the subject of a continuing and vigorous controversy. The many arguments which surrounded the creation and then the more recent abolition of the Land Commission are a case in point. Disagreements are inevitable in judging the point at which the benefits of controls are outweighed by the costs of development delays and other frictions in the working of the economy. However, there would appear to be fairly widespread agreement that the government does have an inescapable commitment to such policy objectives as the prevention of incompatible land uses developing adjacent to each other; the careful valuation of those matters, such as amenity and environmental quality, which the market is unable to measure and is frequently inclined to discount; and the considerable economies which can be derived from efficient transport investments. As a consequence, it simply cannot avoid the supervision of the broad uses to which land is put.

Another area in which public activities inevitably influence the economic and social geography of the country stems from the large and, in recent years increasing, number of people employed in public administration. In 1969, they totalled over 1·4 million, or 6·2 per cent of the country's workforce. Once the sole prerogative of inner London, the location of the central government's civil servants has been considered since 1963 as far from immutable; in the five years to 1969, some 14,000 jobs had been moved away from London to provincial centres. The Post Office, the Inland Revenue and Customs and Excise have been outstandingly 'mobile', the first transferring three of its major sections to Chesterfield, Durham and Glasgow. By 1969, the dispersal of yet a further 26,000 London jobs was under active consideration. Whilst the centrality of London within the national communications system affords it undoubted advantages for a wide range of administrative functions, lower office costs and reduced rates of labour turn-over in the decentralised locations have meant that the need fully to justify any requirements for additional central London office space for government has become more widely accepted in recent years.

Besides its influence over the geography of civil servants, the central government is also inescapably involved in shaping the locational pattern of a wide range of employments in the advanced research and higher educational sectors of the economy, the emerging quaternary sector. The location and expansion of government research establishments, universities and polytechnics are clearly subject to powerful public control and in recent years these institutions have come to be regarded as an important component in inter-regional policies. It is not only local opinion in Teesside, for example, which is prepared to argue that the long-term ambitions of that burgeoning industrial complex make very little sense without at least the establishment of a local polytechnic, and possibly a number of appropriate research establishments as well. Quite apart from the service which such establishments might render local industry, and the innovative enterprises which they might spawn, their (multiplied) employment effects are in themselves capable of making a significant impact upon a region. A university for 4,000 students will provide jobs for approximately 400 teaching staff; these will be assisted by at least an equal number of administrators, library staff, porters, cleaners and others; within the local economy, such a number of people would be serviced by perhaps a further 1,000 employees in schools, hospitals, local authority services, etc.; and the students, too, despite their relatively low purchasing power per head, could well generate locally about 1,000 jobs in shops, public transport, entertainment, the local constabulary and the like. In sum, 4,000 students could well come to be associated with a related population of at least 6,000. In the absence of sub-regional input–output coefficients, the figures in this example are necessarily crude. But the order of magnitude cannot seriously be doubted and underlines the importance which is rightly attached to government policy as it affects the size of universities and other institutions of higher education, especially in the Development Areas.

A further inescapable source of public employment derives from the activities of local authorities. Both the structure of local government and the decisions of local boards – such as hospital management committees – influence not only the patterns of intra-regional service provision but also the geography of employment. Some services, of course, have to be provided in response to the distribution of population; schools and sanitary services are two examples. But other elements in local administration, more especially the higher order services which require a larger population threshold, offer a range of spatial options. The pattern of local government in Britain that will follow from the most recent White Papers on the subject (Secretary of State for the Environment, 1971; Scottish Development Department, 1971; Welsh Office, 1971) will considerably affect the intra-regional geography of public employment. While there will be a marked centralisation of jobs in fewer but larger centres of population, the impact will be smaller than implied by the discarded proposals of Lord Redcliffe-Maud (Royal Commission on Local Government in England, 1969).

Alongside these inescapable commitments of government to shape the economic geography of the country stands the highly ambivalent attitude of government to the nationalised industries. The growing size and importance of public enterprise is inescapable. In the 1965–70 National Plan the nationalised industries (including steel) were to have accounted on average for over 20 per cent of the country's annual investment. But the relationship of government to the behaviour of these industries remains, in the last resort, essentially empirical. It is true that the 1967 White Paper on the *Nationalised Industries* (Ministry of Technology, 1967) spells out more clearly than ever before their economic and financial objectives. This document requires that test rates of discount be used in investment studies and it enjoins the industries not only to cover their accounting costs but to price their goods and services in such a way as to reflect their long-run marginal costs. These are principles which would not embarrass the accountants of a large corporation in the private sector. Simultaneously, however, provision was made for nationalised firms to act against their commercial interests and to be reimbursed from the public purse for doing so. It is all too clear that direct political intervention can influence the investment priorities and decisions of nationalised industries in a way that does not have an exact counterpart in the private sector of the economy. Where a public industry is serving an expanding market, the impact of government intervention in its investment decisions – and hence in the geography of its employment and development – tends to be relatively small. The case of the electricity generating industry, which has been required to burn coal in more power stations than it would have chosen to otherwise and so has stabilised to a small degree the size and location of mining employment, is noted by Manners in Chapter 6. But where the market, and more especially the employment opportunities, of a public industry are tending rapidly to contract, either nationally or regionally, not only is the style of political intervention more varied but that intervention is much more persistent and vigorous.

The local and regional importance of political intervention in the decision-making processes of the nationalised industries cannot be over-stressed. In restraining the British Airport Authority from locating the Third London Airport at Stansted in 1968, the government took note of the number of jobs which would be created within the perimeter of the facility itself, the attraction of industry and the multiplication of these substantial effects through the sub-regional population, as well as the expected transformation of the associated economy. The magnitude of these development implications, and their apparent conflict with the existing strategic proposals for the evolution of land use in the north-east sector of the Outer Metropolitan Area, prompted the government to take the decision out of the hands of the nationalised industry concerned and to offer it to the wider embrace of the Roskill Commission (Commission on the Third London Airport, 1969 and 1970). When the then Minister of Power granted the Central Electricity Generating Board permission to construct a nuclear power station at Hartlepool, he was undoubtedly aware that the decision implied a reduction in the prospective market for coal in the mid-nineteen-seventies by some 3 million tons per year, and thereby jeopardised the jobs of perhaps 2,000 Durham miners. In contrast, by insisting that British Rail should retain passenger services along some of its less travelled routes, Parliament has in effect expressed concern for regional and sub-regional well-being in some of the less densely settled parts of the country, and implicitly made assumptions about at least one of the ways of allocating resources to improve welfare.

Such government intervention in the locational behaviour of the nationalised industries is not inevitable. But it would be most unlikely for it not to occur, given the persistent government involvement in the location decisions of a significant section of private enterprise in recent years. Examples of the latter are provided by the long-standing policies for the distribution of manufacturing and service industry, policies which have sought to reduce inter-regional and even intra-regional differences in the levels of community welfare. Thus, for example, the inducement of the motor assembly industry to move away from the West Midlands and London into Lancashire, South Wales and Scotland; the support of new aluminium smelters in Wales, Scotland and North-East England; the decision to make heavy investments in the infra-structure of Teesside in order to encourage new private enterprise in the region (Secretary of State for Industry, 1963); and the restraints imposed upon office construction in the south-east of England are all part of a calculated public response to a set of regional economic and social problems. They also reflect a particular interpretation of how these problems might be ameliorated.

The deliberate intervention of the government in the spatial allocation of economic activities is the subject of a proliferating literature (Manners, 1972; McCrone, 1969; Brown, 1969). Less fully explored, however, are the spatial implications of policies whose primary aim is non-spatial in character. The Transport Act of 1969, for example, whilst seeking to provide a more logical base for the operation of the national transport system, in fact appears likely to impose a burden upon the rural areas of the country; it certainly embodied a commitment to the maintenance of central government subsidies of the public transport system of Greater London. (This subsidy was, however, removed in 1971.) Again, although the 1967 White Paper on *Fuel Policy* (Ministry of Power, 1967) was primarily designed to balance individual decisions in the market against such national considerations as 'security of supply, the efficient use of resources, the balance of payments and the economic, social and human consequences of changes in the [energy] supply pattern', the details of the policy turned upon the effects of a rapid run-down of the country's coal industry in the higher cost coalfields. The decision to continue with the Concorde programme implies a continuing high level of employment for the people of the Bristol region and a level of prosperity there which might not otherwise have been so easily assured.

Agricultural subsidies have significantly influenced the geography of British agriculture in the past, just as the adoption of import levies on agricultural imports (the practice of the European Economic Community) will undoubtedly influence it in the future; their precise effects upon the patterns of farming in this country, however, are not always easy to untangle from the many other influences moulding farm cropping and employment decisions. The government's decision not to proceed yet with the construction of a Channel Tunnel clearly has implications for at least the timing of certain related developments in south-east and central Kent. Purchase tax lowers the demand for some goods and so reduces the need for labour in some industries – which in turn have their distinctive geography. Import duties in contrast protect the markets of, and employment in, other industries; once again, these have a distinctive spatial expression. And so one could go on. The fact is that virtually all government policies in the end tend to affect the economic and social life of some places either to their advantage or disadvantage. Yet the full spatial effects of these policies are simply never spelled out, even after (let alone before) any new measure is introduced. A balanced interpretation of the widened role of government intervention in the evolution of the country's economic geography, however, cannot afford to discount their importance.

The organisation of local government

The third development which has helped to focus public interest increasingly upon geographical space is in part a reaction to the very power, the widening influence and the undoubted mistakes of the central government machine. The nineteen-sixties certainly witnessed a growing disaffection with the performance and the remoteness of central government, which overshadows an essentially weak and financially hamstrung local government machine. Some see a solution to this problem in a greater 'participation' of more people in all levels of government; the Skeffington Committee (Ministry of Housing and Local Government, 1969), for example, endorsed such a development in the organisation and operation of town and country planning. More universally, however, the reaction has been to question the whole structure of local government and in particular to propose a greater devolution of authority to the regional or provincial scale (Freeman, 1968; Perry, 1969). The process began with the Herbert Commission (Royal Commission on Local Government in Greater London, 1960), which proposed the creation of a new spatial structure for the government of London and initiated the replacement of the London County Council by the territorially larger Greater London Council. The contemporary Boundary Commission for the rest of the country was eventually abandoned, however, in favour of a full-scale Royal Commission on Local Government in England. Its 1969 *Report* advocated the creation of additional GLC-style metropolitan authorities for the conurbations and a larger number of 'unitary' areas to cover the rest of the country. Two years later the government decided to replace the 124 existing English county and county borough authorities by 6 metropolitan areas and 38 administrative counties. Scotland, Wales and Northern Ireland have also witnessed separate investigations of their local government problems, at the same time as influential voices there, if not a majority, pressed for greater political autonomy. Whilst the government also proposed a new framework for local government in Scotland and Wales in 1971, and earlier in 1964 created the office of Secretary of State for Wales (to give the Principality a status similar to that of Scotland), the larger issues raised by the rising tide of Celtic nationalism were handed over to a Constitutional Commission under Lord Crowther for more dispassionate consideration.

In all these debates concerning the spatial organisation of government, two issues repeatedly presented themselves. The first concerned the delimitation of 'areas of common (or community) interest'. At the metropolitan scale, and in the Redcliffe-Maud proposals for 'unitary areas', questions were raised concerning the socio-economic structure of society through space. Should 'x' neighbourhood be placed within 'y' or 'z' borough? At what line on the ground is it most reasonable for metropolitan government to end, given the extending influence of the central areas and the lengthening of commuting patterns? In the case of London, for example, should metropolitan government terminate (approximately) at the inner edge of the green belt, or at the outer edge, or even further from the centre in what is known as the Outer Metropolitan Area? Again, to which unitary area should particular towns most appropriately belong? Does it continue to be an important matter; or are people increasingly indifferent to the *location* of their appropriate town hall, given the increasing spatial flexibility which comes with a greater use of private road transport? What criteria can best be used to determine where the boundary lines should be drawn? Evidence to resolve these and similar issues is by no means conclusive in many parts of the country. Quite frequently efficiency criteria in regard to the performance of local authority services produce very different answers from yardsticks based upon the needs of local democracy and community. And the list of questions extends. Where the most obvious criteria in so many of these matters – indices such as the geography of journeys to work, patterns of shopping behaviour, recreational travel and other simple measures of interaction – cross the historical boundaries of Wales and Scotland, larger issues are implicitly raised (but have never been formally faced) concerning the value of perpetuating indefinitely the political map of the more distant past. The room for argument was, and remains, wide. A central feature of the discussion is the complexity of the problem of changing spatial organisation, a matter upon which central place theory (Dickinson, 1964; Green, 1966) – and, for example, Hall's more empirical discussion of metropolitan growth in Chapter 4 – can throw much light.

The second issue which stemmed from the growing disaffection with the existing structure of government was more strictly economic. The resurgence of Celtic nationalism in the nineteen-sixties, if it was not a response to, certainly fed upon, regional economic dislocation. The response of Whitehall was a sustained attempt to secure larger investments and a stronger flow of British and international funds into these regions. In this narrow respect, policy succeeded remarkably well. Before his resignation in 1969 as Prime Minister of Northern Ireland, Captain T. O'Neill was able to remind – or, in many cases, inform – his fellow Ulstermen that one pound in every three which was spent in the Province originated in London. Mackintosh (1968, 160–1) showed how in 1966/7 United Kingdom government expenditure per head in Scotland was considerably above that in England and Wales for (amongst other items) roads, airports, the promotion of local employment, the support of agriculture, forestry, health and welfare, environmental services, housing and education. Simultaneously, there was a new awareness of the high public costs of maintaining and improving the public transport services of London, of the astronomical cost of the motorway box and other intra-urban road investments being claimed by the Greater London authorities (Thompson, 1969), and of the rising cost of urban redevelopment there. Together, these observations generated a new public interest in the inter-regional flows of public (and private) funds. By 1970, the relevant data with which to examine these spatial transfers were still not available for all the provinces of the United Kingdom. Nevertheless, the search had generated considerable speculation concerning the best criteria for the allocation of funds in the future (Foster and Smith, 1970). Should there be a spatial or regional component? If so, how should it be allocated? According to the size of regional population? According to need, however defined? According to some measure of the return likely to be generated from it?

Preservation of environmental standards

The final development of the last decade which has focussed public interest upon geographical space is the accelerating ease with which the quality of the sub-regional environment can be damaged by uncontrolled or unresearched developments (Arvil, 1967). As technology advances and population expands, so are greater demands made upon the countrys' natural resources of land, water, air and industrial raw materials. As wealth and mobility increase, so does the impact of any individual or group of individuals have a greater geographical extent in the form of larger houses and gardens, the longer journey to work and the ever-extending search for recreational opportunity. Yet simultaneously there emerges the prospect of the land being carelessly utilised, through industrial dereliction or urban blight; of the rivers and the sea being unnecessarily polluted, recent dramatic cases being the pollution of British rivers in 1970 with the closure of sewage works in the course of pay disputes and the previous oil pollution from the wreck of the *Torrey Canyon*; and of the atmosphere steadily deteriorating in some parts of the country through the emission of smoke and fumes by large industrial plants and the automobile. The density of suburban housing is falling, and the problems of pedestrian mobility are tending to increase. Longer journeys to work and the greater use of the private car are creating unprecedented scales and levels of traffic congestion. The search for silence and for some of the more natural styles of recreation is becoming harder. And with these developments has come a wider realisation that wealth and utility cannot be adequately measured by the more traditional yardsticks such as Gross National Product (Mishan, 1967), and that the quality of the sub-regional environment must be more jealously nurtured.

The nineteen-seventies began with quite considerable publicity on the need for public, and sometimes international, action to maintain environmental quality and to check the increase in various forms of pollution. With the 1969 Reith Lectures (Darling, 1970) and the European Conservation Year, the conservationists – strongly reinforced by the preservationists – were given an exceptional opportunity to argue their case. Yet caught somewhat unawares by a sudden public interest in their concerns, the conservation movement was unable to bridge with credibility the yawning gap between the noble claims made for their movement – 'Conservation is a synthesising applied social science crossing the boundaries of all branches of culture... in its observational study of communities' (Darling) – and their somewhat limited achievements in practice, notably the preservation of certain pieces of exceptional countryside and a few items of rare wildlife (Stoddart, 1970). Nevertheless, there is today a juxtaposition of, first, an actual (or threatened) deterioration of environmental quality, which results in particular from the mounting problem of how and where to dispose of waste; second, rising public expectations concerning the quality of local and sub-regional environments, accompanied by a belief that the national wealth is sufficient to allow them to meet their aspirations; and, third, a political awareness of the problem which, with the success of the smoke control legislation behind it, has led to the creation of a permanent Royal Commission on Pollution and the amalgamation of several government departments under a Secretary of State for the Environment. In such circumstances, it is clear that these are matters which will enter increasingly into public debate and will grow rather than diminish in importance.

The quality of public intervention

In any assessment of the quality of public intervention in the processes of spatial allocation, an immediate problem is posed by our ignorance of the full effects of government policies to date and our uncertainty about what the effects of any alternative policies might have been. When in the early nineteen-fifties the government embarked upon a (short-lived) military policy rooted in the so-called independent deterrent, it laid the foundations for the civil programme of nuclear power development. Subsequently some £1,000 million of public funds was invested in the advancement of nuclear technology. An increasing number of nuclear power stations – the economics of which are by no means free from debate – has subsequently appeared on the British landscape. That they have immediately displaced large tonnages of coal and presented considerable problems to the higher-cost coalfields cannot be denied. However, what is less certain is whether, without the development of British nuclear technology, the coal industry and the coalfield communities would have been in a better or worse position, since it is impossible to know what would have been the government's attitude to the other and certainly cheaper alternative, oil. Similarly, the government's decision to inject large sums of public money into the (private) cotton industry, under the Cotton Industry Act of 1959, in order to assist in its rationalisation, substantially modified the geography of textile employment in the United Kingdom; but it is impossible to specify with any certainty how different would have been the evolution of the industry had it been left to face the rising tide of foreign competition solely from its own resources. Again, there are clearly long-term implications for the future economic geography of the country stemming from the government's decisions to approve new deep-sea container terminals at Tilbury, Liverpool, Southampton and – just before the 1967 Pollok bye-election – Glasgow. Their exact nature, however, remains unexplored (Tanner and Williams, 1967). Even more uncertain are the consequences which followed from the Ministry of Transport's refusal in 1966 to endorse the city of

Bristol's scheme for an entirely new container and bulk commodity facility at Portbury (Ministry of Transport, 1966) and the 1970 reversal of that decision by the new Conservative Administration.

By the same token, whilst it is possible to point to the London new towns as an example of deliberate intra-regional location policy, it remains very difficult to judge the *net* effect of government action when it is realised that only one in eight of the population moving out of Greater London in the post-war years was destined for these and other *planned* reception centres. Keeble (1972) has pointed to the powerful centrifugal forces shaping the location of economic activities and people in South-East England, and has stressed the accord between natural tendencies and the government's programme of overspill. As with the control of outdoor advertising, which is one of the outstanding achievements of British environmental planning in the last quarter century, it is impossible to measure reality against a hypothetical model of what might have been.

Any assessment of the public record in the matter of spatial allocation is faced with a second fundamental problem. This is the crudeness and vagueness of government remits and goals. For many years now, successive Acts of Parliament relating to the inter-regional balance of economic activities have empowered government to pursue 'the proper distribution of industry'. Satisfactory though the phrase might be to the legislators and lawyers, in fact we are still no clearer about what *is* the 'proper' distribution of industry than we were thirty years ago. Again, planning remits have frequently invoked the pejorative word 'congestion', without first seeking accurately to examine its meaning; to distinguish it from 'concentration', and to explore whether or not – and at what point – it appears to generate net disbenefits for the individual, society or the government, is a task still waiting to be accomplished. The green belt vision of Abercrombie, so persistently defended by successive Ministers of Housing and Local Government, was very different from the complex reality described by Thomas (1970). The 'growth points' and 'growth zones' of 1963 vintage (Secretary of State for Industry, 1963; Secretary of State for Scotland, 1963) may have captured the imagination of Lord Hailsham and the Scottish Office, but their delimitation eluded precise definition.

Thus, with even the formal spatial policies of governments frequently having imprecise goals, and with the secondary effects of non-spatial policies playing a considerable role in the evolution of the country's economic geography, judgements concerning the quality of public intervention in the processes of spatial allocation must inevitably be highly qualified. Four points can, however, be made with reasonable confidence.

The first point is that the processes of spatial allocation, and hence the causes of spatial dilemmas, arc coming to be better understood. There was a time when it was the symptoms of spatial problems which generated a political response, whilst the underlying causes were left unaddressed. The initial response to the dilemmas of the older industrial areas, as embodied in the Special Areas legislation of the inter-war years and the post-war Distribution of Industry Acts, was to invoke measures designed to ameliorate the principal symptom of the problems there – unemployment. Yet measures designed specifically to increase employment frequently left unaltered the fundamental causes of a low level of economic activity. However, by the time of the Industrial Development Act of 1966, and more especially with the Report of the Hunt Committee (Secretary of State for Economic Affairs, 1969), it was evident that a much improved understanding of the causes of this inter-regional problem was emerging. This in turn generated a more varied and better coordinated set of objectives and policies for the Development and Intermediate Areas, policies better attuned to assist in their long-term recovery. This is not to deny that many fundamental dilemmas remain. Warren in Chapter 7 discusses some of the unresolved questions facing the chemical industry as it experiences new locational pulls. Keeble exposes some of the fundamental constraints upon the mobility of employment in Chapter 2. Nevertheless, there is no denying that the understanding of inter-regional growth and its associated dilemmas has advanced substantially during the last decade.

Similarly, research into such matters as the behavioural aspects of industrial and office location decisions, the processes of urban growth, the supply of and the demand for recreational land uses and the inter-relationship between traffic and land use has helped to shape more effective policies for the creation of a satisfactory sub-regional environment. Although this is an area of public activity which is highly subject to the swings of fashion in physical planning and architecture, there can be little doubt that the intellectual underpinnings of, say, *Teesplan* (Teesside Survey and Plan, 1969) or the Leicestershire Sub-Regional Plan (McLoughlin, 1969) are much firmer than equivalent documents a decade or so earlier.

Related to this improved understanding of the causes of regional and sub-regional problems has been the development of a new cluster of techniques of spatial analysis and investment appraisal. The refinement of economic base and multiplier theory, the development of inter-industry and industrial complex studies, the improvement of linear programming and systems analysis, and the exploitation of cost–benefit studies have together considerably improved the analytical and predictive tools available

2

to those engaged in decision-making. In Chapter 3, for example, Haggett discusses the value of lead–lag information on one economic indicator in the specification of regional forecasting models and regional systems analysis. Without doubt the most ambitious and costly of the tasks which public authority set itself using such 'modern' techniques was that attempted by the Roskill Commission on the Third London Airport (Commission on the Third London Airport, 1969 and 1970). Forecasting techniques were used to arrive at estimates of the expected level of aircraft and air passenger movements to the year 2001; economic base and multiplier analyses were needed to produce data on the size and the structure of the associated employment and urbanisation; a gravity model was used to allocate passengers to their (inland) origins and destinations in relation to the airports of the London system; an accessibility model was applied to reach conclusions concerning generated traffic; and a most ambitious cost–benefit exercise sought to relate all these findings to a common set of (discounted) values. The criticisms of the techniques were, predictably, abundant (Adams, 1970; Self, 1970). But comparison of the procedures adopted to decide the location of the first (Heathrow) and second (Gatwick) London airports, and what might have been the third (Stansted), with those used by the Roskill Commission convincingly suggest, if nothing else, a marked improvement in our understanding of the issues involved. At the very least, it cannot be denied that with these techniques the area of uncertainty in decision-making is reduced, and the matters more legitimately left for political appraisal are more clearly exposed. In the rare instance, it might even be possible to claim with some certainty that a right decision has been indicated and espoused.

Moreover, it is important to recognise that the administrative spatial framework for public activity has been substantially improved in recent years. There is little doubt now that the organisational base that was provided by the 1947 Town and Country Planning Act, which vested power in the county and county borough councils for the designation of land-use plans, was less than perfect. As the mobility of people has increased and as the extent of their spatial impact has widened, that geographical basis for decision making has become weaker still. In addition, although one ministry was given responsibility for land-use planning, other government departments were given charge over matters which were intimately related to the spatial allocation process. The Board of Trade supervised policies relating to the distribution of industry; the Ministry of Transport, with its responsibility for supervising investment in transport facilities, had an important impact upon spatial relationships; and the Ministry of Labour (now the Department of Employment) bore responsibility for matters concerning labour retraining. With the creation of a Minister for Regional Development in 1963 – a post which had only a brief life, but which was given for the first time the task of attempting to coordinate inter-departmental activities relating to regional policy – a new pattern began to emerge. In 1964/5 the regional planning boards and councils were created under the Department of Economic Affairs. Then, with the appointment in 1969 of a Secretary of State for Regional Planning and Local Government (renamed the Secretary of State for the Environment in 1970 and given a slightly modified remit), a single office was given the responsibility of overseeing most matters which affect inter- and intra-regional planning and included the Ministries of Local Government and Development, Housing and Construction and Transport. Although policies affecting the location of manufacturing industry were paradoxically left with the Ministry of Trade and Industry, the government clearly recognised in these changes the need for a new framework within which to consider spatial planning matters. A parallel development in local government was the creation of a number of Standing Committees on regional planning, through which adjacent counties and county borough councils could coordinate their planning strategies and actions. The Standing Conference on London and the South East was the first in the field, but it was quickly followed by a number of others established on either an *ad hoc* or a permanent basis.

The proposals of the government in 1971 for local government reform, together with the Crowther Report on the Constitution which is still awaited at the time of writing, prospectively will throw the geographical basis of physical and regional planning into a fluid state once again. However, there can be little doubt that the difficulties imposed by the spatial administrative structure containing these activities in the past are in the process of being reduced. A much greater awareness exists today than ever before of the need to provide as appropriate a framework as possible for public intervention in the evolution of the country's economic and social geography.

Recently, then, the causes of regional and sub-regional problems have come to be better understood. The techniques of spatial analysis and of investment appraisal have assumed a greater sophistication. And a more relevant geographical framework for the administration of public decisions has begun to emerge. Nevertheless, weaknesses and inconsistencies in public policies relating to geographical space remain. Quite apart from the high costs of delays in decision-making, which characterise sub-regional planning policies in particular, and occasional direct intervention of party and regional politics into major investment decisions (the location of the steel strip mills in Wales and Scotland in 1958, and the siting of the aluminium smelters ten years later, are two cases in point), a number of more fundamental dilemmas remain. The priorities of resource allocation, and the specification of clear planning goals, continue to be singularly elusive. The most appropriate modes of intervention in the space transforming processes remain more intuitive than clear. As a consequence, the constantly changing meaning of space in economic and social life affords a set of continuing dilemmas for public policy-making.

The size and the importance of these questions has naturally attracted the attention of workers in a variety of academic disciplines. Economists for the first time have recently made a major excursion into the field of regional economics (Brown, 1969; Needleman, 1968; Richardson, 1970) and have joined vigorously in the debate on inter-regional development policy. Sociologists and political scientists have both come to a new awareness of geographical diversity and spatial organisation (Pahl, 1968; Tress, 1967, 1969). But it has been geographers who, albeit partly handicapped by the descriptive traditions of their discipline, have emerged from the last decade with the most acute awareness of the contribution which their enquiries can make to the clarification and the solution of a widening range of public issues. Something of this awareness and of this potential is reflected in the chapters which follow.

Contributions to the debate

Problems relating to the spatial allocation of resources are clearly manifest at all geographical scales, from the local issues that worry small communities through problems at the intra-urban scale to questions of inter-regional balance. The contributors to this volume have pitched their discussion at the second and third levels. Pahl, concerned with the distribution of incomes and of access to facilities, discusses these issues in the context of the built environment of large urban areas; Haggett has chosen a similar scale, concentrating on the intra-regional patterns of variation in the level of employment and using data for the South West region. By contrast, Hall's analysis of urban growth and development ranges from the conurbation level to the national distribution of urbanisation, whilst the other contributors for the most part all couch their discussion in terms of inter-regional and national issues.

In temporal terms, the contributions again cover a wide range. Haggett's concentration on short-term leads and lags in employment levels, measured in months, is atypical of the contributions. Keeble's concern with the spatial mobility of employment has a relevance in terms of changes over, say, half a decade and longer. A similar time scale applies to Warren's discussion of the changing balance of location factors in the chemical industry and to Manners' review of competition between energy sources in a national market that is spatially differentiated, giving some fuels a greater relative advantage in some locations than in others. On the other hand, Hall's examination of urban growth takes a longer historical view than any other contribution and involves a time scale measured in decades. It is this order of time that is also relevant to Chisholm's evidence concerning the spatial variation in the cost of transport as it affects the location of employment.

A third feature of the volume deserves to be noted. There is no sector of the economy or section of society for which allocations in space are not important in some degree. To this extent, therefore, the problems posed are of general application. However, with the exception of Pahl's chapter, all the contributions deal with 'economic' rather than 'social' aspects of the space-economy. Within this general heading of 'economic' phenomena, Hall's concern is with the distribution of population and the associated built forms, while Haggett and Keeble deal with questions of employment. On the other hand, particular industries are faced with their own problems, as discussed by Warren for chemicals, Manners for the energy industries and Chisholm for the impact of transport costs on all activities.

The chapters in this volume, being limited in number, can only consider a small number of topics from a very large field. Within the three dimensions noted above – space, time and sector – the contributions are quite widely distributed. They serve, therefore, not to give a complete account of the relevance of geographical space in the making of policy decisions but rather to demonstrate the wide range of issues for which a geographical contribution has a relevance. All the chapters are original pieces of work and represent a substantive contribution to an important debate.

References

Adams, J. (1970). 'Westminster: the fourth London Airport', *Area*, 2, 1–9.
Arvil, R. (1967). *Man and Environment*, Penguin.
Brown, A. J. (1969). 'Surveys in applied economics: regional economics, with special reference to the United Kingdom', *The Economic Journal*, LXXIX, 316, 759–96.
Central Unit for Environmental Planning (1969). *Humberside: a Feasibility Study*, HMSO.

Chisholm, M. (1970). *Geography and Economics*, Bell, 2nd rev. edition.

Commission on the Third London Airport (1969 and 1970). *Papers and Proceedings*, HMSO.

Darling, F. F. (1970). *Wilderness and Plenty*, BBC.

Department of Economic Affairs (1967). *The Development Areas: a Proposal for a Regional Employment Premium*, HMSO.

Dickinson, R. E. (1964). *City and Region: a Geographical Interpretation*, Routledge.

First Secretary of State and Secretary of State for Economic Affairs (1967). *The development areas – regional employment premium* (Cmnd. 3310), HMSO.

Foster, C. D. and J. F. Smith (1970). 'Allocation of central government budgets over city regions', *Urban Studies*, VI, 2, 210–26.

Freeman, T. W. (1968). *Geography and Regional Administration*, Hutchinson.

Green, F. H. W. (1966). 'Urban hinterlands: fifteen years on', *Geographical Journal*, CXXXII, 263–86.

Keeble, D. (1972). 'The South East and East Anglia', Chapter 2 in G. Manners (ed.).

McCrone, G. (1969). *Regional Policy in Britain*, Allen and Unwin.

Mackintosh, J. P. (1968). *The Devolution of Power*, Penguin.

McLoughlin, J. B. (ed.) (1969). *Leicester and Leicestershire Sub-Regional Planning Study*, Leicester, 2 vols.

Manners, G. (1966). *The Severn Bridge and the Future*, TWW Cardiff.

Manners, G. (ed.) (1972). *Regional Development in Britain*, Wiley.

Ministry of Housing and Local Government (1969). *People and Planning* (Report of a Committee under the Chairmanship of A. M. Skeffington), HMSO.

Ministry of Power (1967). *Fuel Policy* (Cmnd. 3438), HMSO.

Ministry of Technology (1967). *Nationalised Industries – A Review of Economic and Financial Objectives* (Cmnd. 3437), HMSO.

Ministry of Transport (1963). *Traffic in Towns* (Reports of the Steering Group and Working Group), HMSO.

Ministry of Transport (1966). *Portbury. Reasons for the Minister's Decision not to Authorise the Construction of a New Dock at Portbury, Bristol*, HMSO.

Ministry of Transport (1969). *Roads for the Future – a New Inter-urban Network*, HMSO.

Mishan, E. J. (1967). *The Costs of Economic Growth*, Staples.

Needleman, L. (ed.) (1968). *Regional Analysis*, Penguin.

Pahl, R. E. (1968). *Spatial Structure and Social Structure*, Centre for Environmental Studies, Research paper.

Perry, N. H. (1969). 'Geography and local government reform', Chapter 22 in R. U. Cooke and J. H. Johnson (eds.), *Trends in Geography, an Introductory Survey*, Pergamon.

Richardson, H. W. (1970). *Regional Economics: A Reader*, Macmillan.

Royal Commission on Local Government in England (1969). *Report* (Chairman Lord Redcliffe-Maud, Cmnd. 4040), HMSO.

Royal Commission on Local Government in Greater London (1960). *Report* (Chairman Sir Edwin Herbert, Cmnd. 1164), HMSO.

Scottish Development Department (1971). *Reform of Local Government in Scotland* (Cmnd. 4583), HMSO.

Secretary of State for Economic Affairs (1969). *The Intermediate Areas* (Report of a Committee under the Chairmanship of Sir Joseph Hunt, Cmnd. 3998), HMSO.

Secretary of State for the Environment (1971). *Local Government in England* (Cmnd. 4584), IIMSO.

Secretary of State for Industry, Trade and Regional Development (1963). *The North East – a Programme for Regional Development and Growth* (Cmnd. 2206), HMSO.

Secretary of State for Scotland (1963). *Central Scotland – a Programme for Development and Growth* (Cmnd. 2188), HMSO.

Self, P. (1970). 'Nonsense on stilts: the futility of Roskill', *New Society*, July 2, 8–11.

Stoddart, D. R. (1970). 'Our environment', *Area*, 1, 1–4.

Tanner, M. F. and A. F. Williams (1967). 'Port development and national planning strategy: the implications of the Portbury decision', *Journal of Transport Economics and Policy*, I, 3, 1–10.

Thomas, D. (1970). *London's Green Belt*, Faber and Faber.

Thompson, M. J. (1969). *Motorways in London*, Duckworth.

Teesside Survey and Plan (1969), *Teesplan*, HMSO.

Tress, R. S. (1967). 'The new regional planning machinery and its research needs', *Regional Studies*, I, 1, 23–6.

Tress, R. S. (1969). 'The next stage in regional planning', *Three Banks review*, 81, 3–30.

Welsh Office (1971). *The Reform of Local Government in Wales*, HMSO.

2. Employment mobility in Britain

DAVID KEEBLE

In many ways, the most fundamental, striking and effective aspect of post-war government policies aimed at redressing inter-regional economic inequalities in Britain has been the promotion of industrial movement from the prosperous central regions to the lagging peripheral regions of the country. At the same time, what few coherent intra-regional planning policies are being implemented, notably for such areas as south east England, the Glasgow region, and the west Midlands, depend essentially upon the physical transfer of employment from congested central conurbations or cities to outlying settlements, including such famous new towns as Stevenage, Harlow, Cumbernauld and the incipient Milton Keynes. At both scales, national and regional, employment mobility is thus of crucial importance for successful implementation of government physical and economic planning policies, as well as for any prediction of the future configuration of the country's space-economy.

The validity of these assertions is evident not least from the spate of government legislation concerned with the promotion of employment mobility which has been enacted since 1945. Stimulated initially by the Barlow Commission Report on the Distribution of the Industrial Population (Royal Commission, 1940), and encouraged increasingly by political pressures from such peripheral areas as Scotland and Wales, successive post-war governments have placed on the statute book at least eight major Acts concerned wholly or in part with encouraging the movement of economic activity within Britain (Dowie, 1968; McCrone, 1969). The Distribution of Industry Act 1945, the Distribution of Industry (Industrial Finance) Act 1958, the Local Employment Acts of 1960 and 1963, the Finance Acts of 1963 and 1966, the Control of Office and Industrial Development Act 1965, and the Industrial Development Act 1966, bear witness to the importance attached by politicians to promoting employment mobility on the inter-regional scale. So too does the associated and now familiar jargon, with its references to IDCs (Industrial Development Certificates), ODPs (Office Development Permits), REP (Regional Employment Premium), and Development, Special Development and Intermediate (or 'Grey') Areas. The last of these were of course the subject of a special enquiry (Secretary of State for Economic Affairs, 1969), whose largely abortive recommendations were fundamentally concerned with modifying to some slight degree current industrial movement patterns.

As an index of the importance of employment mobility for national and regional planning in Britain, government concern is moreover matched by the remarkable scale of post-war movement, at least of manufacturing industry. In 1968, the Board of Trade published for the first time (Howard, 1968) estimates of the volume of manufacturing movement across regional boundaries in the United Kingdom between 1945 and 1965. This study revealed that according to the Board of Trade's records, the country's peripheral areas (Scotland, Wales, the Northern region, Northern Ireland, Merseyside/South-East Lancashire, and Devon and Cornwall) received from other regions of Britain during this period no less than 921 manufacturing establishments, which by the end of 1966 were providing jobs for 345,000 workers. Migrant firms from other regions were thus responsible by 1966 for as much as 25.9 per cent (Wales) and 18.5 per cent (Northern region) of total manufacturing employment in particular peripheral areas. Spooner's independent survey (1970) of Devon and Cornwall suggests that the figure there may have been even higher, of the order of 27.5 per cent in 1965 (for movement after 1939). These statistics indicate that the contribution of industrial movement to the provision of employment in the lagging peripheral regions has been a significant one. Moreover, intra-regional mobility, essential for implementation of the various regional new town schemes, has also been considerable. Most notably, manufacturing movement from the densely built-up GLC area outwards to the rest of South-East England and East Anglia involved the establishment by 1966 of at least 825 factories, and the provision of 194,000 jobs. This Board of Trade migration employment estimate represents approximately 15 per cent of total 1966 manufacturing employment in these reception areas. An even more recent study (South East Joint Planning Team, 1970), defining movement so as to include transfers to a new address over any distance, even of a few metres, revealed that a much higher percentage (nearly half) of all South East manufacturing units employing over 100 workers in 1968 had shifted their location in some way within the region since 1945. The scale of industrial mobility in post-war Britain thus seems often to have been considerable, a finding which contrasts with Smith's conclusion (1954, 49) from American studies that manufacturing moves account for 'but a small part of the overall change in manufacturing importance of an area'.

The significance of employment mobility from a planning viewpoint is further enhanced by the fact that, again at least as far as manufacturing industry is concerned, post-war movement, far from embracing a cross-section of all industry, has been highly selective of the country's biggest, most rapidly expanding and most export-orientated firms. Though no data on firm size are presented by Howard, a 1964 outer North-West London survey (Keeble, 1968, 20–3), for example, established conclusively that for this major industrial area, outward manufacturing movement over distances of 16 kilometres (10 miles) or more has been dominated by the area's larger firms, which have participated in migration to a much greater degree than their numbers in the area would lead one to expect. The converse is true of the area's smaller firms. This finding undoubtedly applies to movement from Greater London as a whole, and probably also to that from the West Midlands conurbation and other large industrial centres. Its expression in reception areas is well illustrated by the giant motor-car assembly plants established during the early nineteen-sixties by Fords, Vauxhall and Standard–Triumph (now British Leyland) on Merseyside, and by Rootes (now Chrysler) at Linwood in Scotland. Ford's Halewood plant alone cost over £30 million and now employs over 13,000 workers. Though detailed size data are not available, and dispersal has been much more restricted spatially, it seems very probable that large office firms also account for an unusually large share of office movement from central London, relative to their importance there.

That manufacturing movement has been highly selective of the country's more virile firms and leading growth industries has been well established by numerous recent detailed surveys of particular groups of migrant firms (Cameron and Clark, 1966; Loasby, 1967; Keeble, 1968; Spooner, 1970). Howard's data permit further corroboration of this association at a much more aggregate level. Figures from Table 9 of his study (Howard, 1968, 27), relating to industrial categories defined at the detailed Minimum List Heading level, so as to distinguish as clearly as possible between growing and contracting manufacturing activities, are plotted in Figure 2.1. Along the X-axis are recorded estimates of United Kingdom employment growth or decline in each industry between mid-1953 and November 1966, while along the Y-axis are recorded estimates of the amount of employment created by the end of 1966 by migrant firms which moved between 1952 and 1965, again by industrial category. The most important conclusion to be drawn from the graph – and one stressed by Howard in his discussion of the raw data – is that in industry terms, volume of growth and volume of movement are very closely related. Correlation of these two variables for the population of expanding industries ($N = 11$) yielded an r value of 0.9441 ($r^2 = 0.8913$), which is significantly different from zero at the 0.001

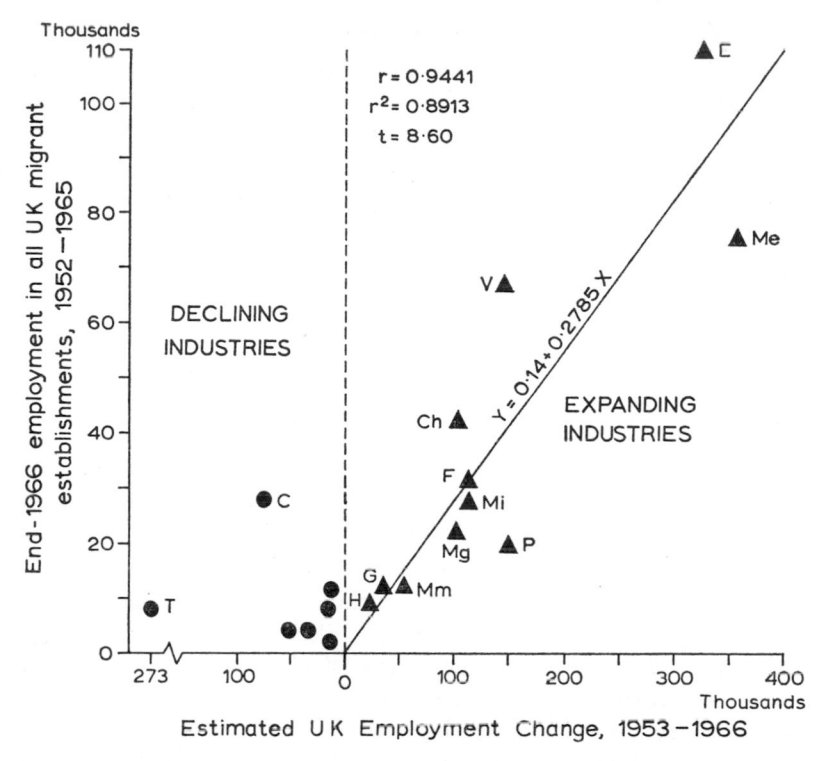

Figure 2.1. United Kingdom: manufacturing mobility and growth, 1952–66.
C: Clothing and footwear. Ch: Selected chemicals and man-made fibres.
E: Electrical goods and scientific instruments. F: Selected food, drink and tobacco
industries. G: Glass, cement, abrasives etc. H: Hosiery, carpets and 'other
textiles'. Me: Selected mechanical engineering. Mg: Selected metal goods.
Mi: Selected miscellaneous industries. Mm: Metal manufacture (exc. castings).
P: Paper, printing and publishing. T: Selected textiles. V: Motor vehicles and
aircraft.
Source: Howard, 1968, Table 9.

level. In other words, 89 per cent of the variation in mobile employment
values is 'explained' statistically by variation in volume of growth. The
findings of local area 'micro-level' studies, that variations in manufacturing
movement are chiefly to be explained by inter-industry growth rate
differences consequent upon different national and/or international trends
in demand for their products, are thus fully substantiated.

Indeed, the level of statistical explanation demonstrated is perhaps
surprisingly high, given the approximate nature of the employment growth
estimates, and the further stress in micro-level studies upon the importance

for inter-industry movement rates of differences in the 'potential mobility' of firms and industries (see Keeble, 1968, 14–19). Potential mobility here refers to the relative intensity of local bonds tying an existing firm or industry to its current location, usually an urban area. The bonds envisaged are those which inhibit movement over distances of more than a few kilometres, and include orientation to a specialised, perhaps highly-skilled, labourforce, dependence upon local customers or raw materials, the need for access to specialised transport facilities, and enjoyment of local 'juxta-position economies' (Stevens and Brackett, 1967, 2), such as those which may arise from local information, material or processing linkages. Though extremely difficult to quantify (see below), inter-industry variations in potential mobility almost certainly account for the bulk of 'unexplained' variation revealed by deviations of industry points in Figure 2.1 from the best-fit regression line. Thus the three industries – electrical goods, motor vehicles and aircraft, and selected chemicals – exhibiting higher movement rates than predicted are all unusually 'footloose', in the sense that produc-tion technology, labour requirements and marketing considerations do not greatly inhibit movement, at least within the 'effective area' of the country (Caesar, 1964, 230). Greater potential mobility here is also undoubtedly related to the relatively large average size of plant in these industries. In contrast, as micro-level studies have shown (Keeble, 1968, 17–19), many firms in the mechanical engineering and paper, printing and publishing industries are small and potentially less mobile, tied to large centres such as London and Birmingham by a web of dependency relationships with local suppliers, customers and skilled workers. For a given expanding industry, the marked stimulus to movement afforded by growth and consequent factory space and labour shortages is thus moderated or enhanced by the further variable of potential mobility.

Of course, as Figure 2.1 demonstrates, not all movement is associated with expansion. The clothing and footwear trades, forced to migrate by rising labour costs and the need to tap local supplies of relatively cheap female labour, stand out as a major anomaly in this respect. The textile industry, employment in which declined by the staggering total of 273,000 workers between 1953 and 1966, also contributed to migration. Growing agglomeration diseconomies such as rising labour and factory rental costs in traditional urban–industrial centres for contracting industries reliant upon cheap female labour probably lie behind this phenomenon. It should not obscure the fact, however, that no less than 86 per cent of all movement during the 1952–65 period, measured by employment created, was in expanding industries. It is extremely interesting to note the remarkable similarity of this percentage, derived from aggregate statistics, with that obtained by micro-level studies in a similar context. Thus entirely independent surveys by the present author (Keeble, 1968, 4), Cameron and Clark (1966, 74) and Spooner (1970) yielded values of 85 per cent, 85 per cent and 83 per cent, respectively, for the proportions of the particular sample of migrant establishments studied which had been set up as a direct consequence of expansion. Only Townroe's study (1971), which was based on a smaller sample than any of the others, differs slightly in this respect (73 per cent). The striking uniformity of the first three 'large-sample' results, and their similarity with the aggregate percentage derived from Board of Trade data, strongly suggest that a value of approximately 85 per cent represents a kind of equilibrium level which defines the relative contributions of expanding and contracting industry in any reasonably large sample of post-war migrant British manufacturing firms.

One of the more interesting findings of the North-West London study already mentioned (Keeble, 1968, 34–5) was its conclusion concerning the selectivity of migrant firms with regard to their level of export activity. No other published surveys refer to this point in detail. The North-West London study, which investigated a sample of 65 firms that had transferred entire production over distances of at least 16 kilometres (10 miles) since 1940, discovered that 84 per cent of these migrant concerns were exporters. The corresponding proportion of a randomly-selected sample of resident firms ($N = 124$) in the origin area was only 65 per cent. For firms exporting one-quarter or more of total production, the respective proportions were 33 per cent (migrant) and only 17 per cent (resident). The χ^2 test reveals that this latter difference is highly significant (at the 0.001 level). At least in the case of manufacturing movement from London, migrant firms are significantly differentiated from the total population of firms in terms of much greater export orientation.

In all these ways, therefore, the movement of manufacturing activity within Britain is more important for national and regional economic planning than even its considerable volume would suggest. Firms which move tend to be significantly bigger, more virile and more export-orientated than the population from which they are drawn. Such characteristics not only enhance the contribution which mobile firms make to the economies and employment prospects of reception areas, but indicate the need for great care on the part of government in framing and implementing employment mobility policies, if national economic prosperity and growth are to be safeguarded.

Movement definition

Although both micro-level studies and Howard's data are beginning to permit some evaluation of the scale and character of post-war employment mobility, many questions at present remain unanswered. At a very simple factual level, for example, and despite the activities of both the Location of Offices Bureau (1970) and academic observers (Wabe, 1966; Daniels, 1969), insufficient is yet known about the volume and pattern of office mobility within Britain, although stringent office building controls apparently designed to promote inter-regional office movement have been in force since November 1964. Recent movement estimates, such as Hall's calculation for the Location of Offices Bureau (LOB) of an annual rate of office decentralisation from central London of 30,000 jobs per annum, mid-1963 to late-1969, are perforce based on samples, and hence somewhat tentative (Hall, 1970). The scale of movement of other services, such as warehousing, education and health facilities, is almost totally unrecorded. Admittedly, movement of such activities occurs chiefly on an intra-urban scale. But examples of migration over reasonably long distances, as with the shift of the University of Surrey from Battersea to Guildford, or the part-transfer of Sainsbury's wholesaling activity from Blackfriars to Basingstoke (Hampshire) and Buntingford (Hertfordshire), seem to have become sufficiently common to warrant investigation.

However, perhaps even more important are unanswered but fundamental questions concerning the precise definition of employment mobility, the accurate identification for planning purposes of mobile and immobile firms, and the best ways of evaluating the reasons for variations in movement streams between different parts of the country. The problem of movement definition, apparent not least from the very different approaches adopted by different recent mobility studies, has at least four aspects. Least complicated is the definition of the discrete economic units whose movement is to be studied or channelled. Both planning action and academic studies have almost invariably focussed on the establishment as the fundamental, 'first-order', object of interest in this context. In the manufacturing case, this is represented by a separate factory, with its workforce and plant: in the office case, by a separate office department or unit, again with its workforce. The important distinction is thus between establishment and firm, the latter of which may operate several establishments.

More complex is the distance aspect of movement definition. The problem here is the lack of a commonly accepted distance definition as between different studies. Logically, the setting-up by an existing firm of any new or

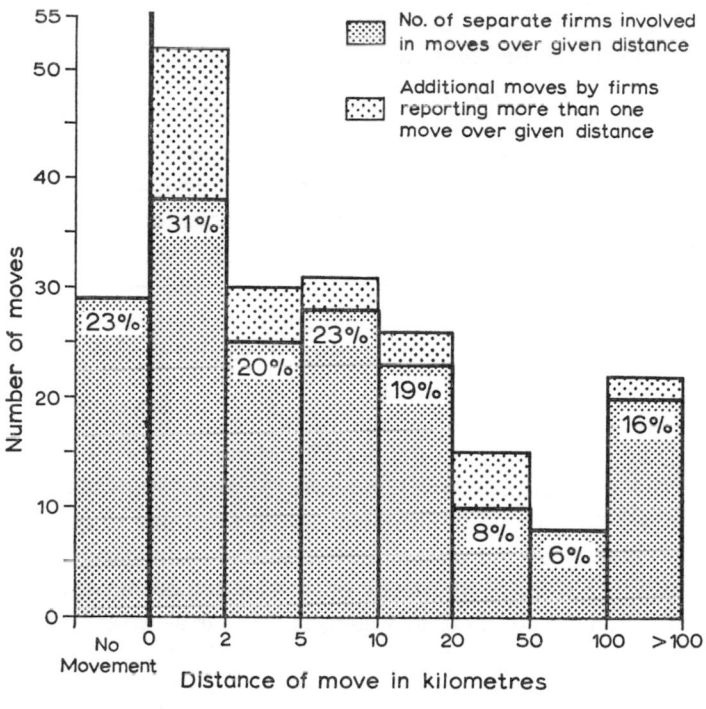

Figure 2.2. North West London: distance and movement frequency of manufacturing firms.

Note: The percentage value in each column refers to the proportion of the total sample of firms (124) reporting at least one move over the distance range specified. The first column refers to the number and percentage of firms reporting no moves since manufacturing began in Britain.

Source: N.W. London industrial location survey, unpublished data.

replacement establishment on a site which is not actually part of or contiguous with that already occupied by the firm concerned involves movement. However, for manufacturing at least, this definition embraces a very large number of 'moves' over only very short distances. This is clearly illustrated by Figure 2.2 which is compiled from unpublished data obtained from a 1963 interview survey of 124 randomly selected manufacturing firms employing between 10 and 999 production operatives in outer North-West London. The survey revealed that if movement is defined as above, no less than 77 per cent (95) of the firms interviewed had been involved in at least one move since manufacturing began in Britain (several firms also had some sort of overseas origin). Altogether, the sample reported no less than 184 moves, or 1.94 moves for each of the 95 firms reporting movement. However, the important point here is that no less than 52 of the 184 moves (28 per cent) were over distances of less than 2 kilometres (1.24 miles) while 113 moves (61 per cent) were over distances of less than 10 kilometres (6.2 miles).

This very high frequency of short-distance movement, which involves both complete relocations and occupancy of additional local branch factories, has generally been ignored by most studies of manufacturing migration in Britain. A major exception is Cameron's current research project on manufacturing shifts within Glasgow, building on an earlier more limited study (Cameron and Johnson, 1969). Admittedly, intra-urban migration is probably less significant from a planning viewpoint than longer-distance movement, which may be channelled by the planner to needy reception areas such as the peripheral regions or new towns (Smith, 1970, 193–4). It is interesting to note in Figure 2.2 the surprisingly high frequency (22) of moves of over 100 kilometres (62 miles), the bulk of which represent branch factory establishment or forthcoming long-distance transfer from North-West London of the firm's entire manufacturing activity. However, the high frequency of short-distance moves, and the survey's results as a whole, do clearly suggest that intra-urban movement, with its implications for local journey-to-work patterns and policy, deserves more attention than it has hitherto received; and that any minimum distance definition of 'movement' is bound to be arbitrary, at least in terms of the graph of movement frequency against distance. In other words, any assessment of the importance or scale of manufacturing movement in a particular study dealing with a particular area is largely meaningless without clear specification of the distance definition adopted. The significant differences in practice are illustrated by the distance definitions used by Howard (1968) – movement across any boundary defining one of 50 areas in the United Kingdom, involving in general distances of 32 kilometres (20 miles) and upward; by the present author (Keeble, 1968) – movement to a new location 16 kilometres (10 miles) or more from the former or main factory; and by Economic Consultants Ltd in their survey for the South East Joint Planning Team's report (1970) – movement over any distance, as in Figure 2.2. Though the discussion here has concerned manufacturing industry, exactly the same considerations apply in the case of office and service industry mobility.

A third definitional problem arises from the common adoption of a firm size threshold below which an establishment, however mobile, is omitted from analysis. In fact, of course, small size is fairly highly correlated with length of move. Thus despite the use of 1963 size values (rather than size values relating to date of move), the North-West London sample survey, which concerned only firms employing at least 10 production operatives, revealed that the average size of firms involved in moves of less than 10 kilometres was 148 employees compared with a value of 261 employees for firms moving over 10 kilometres. Exclusion of firms employing less than, say, ten employees (Howard, 1968), or as an even higher threshold, one hundred employees (Economic Consultants Ltd), may thus automatically exclude from consideration much short-distance movement. Whether or not this is important depends on the purpose of the study concerned.

The final problem of definition involves the organisational aspect of movement. All mobility studies, whether of manufacturing or service industry, accept that movement can take at least two forms. One is the complete relocation, or wholesale transfer, of a firm's existing economic activity from one location to another. The other is movement through the establishment in a new location of a branch unit of the firm concerned, which nonetheless maintains its activity in its existing location. Very few studies, of which that by Economic Consultants Ltd is a major and notable exception, define movement to include the establishment of entirely new enterprises. However, as Howard (1968, 3–5) points out, in a modern industrial economy, locational shifts sometimes defy easy classification into 'moves' and 'non-moves' on the basis of these simple definitions. The most important and complex examples result from take-overs or mergers with other existing firms, and the concomitant development of large multi-plant corporations. Take-overs, which have been increasingly frequent in recent years, almost always result in rationalisation and re-assessment of the locational pattern of the enlarged firm's economic activity. As a result, establishments may cease, significantly reduce, or expand their activities, while in the manufacturing case, plant and personnel may be transferred from one location to another. The impact of such locational shifts upon particular areas may be considerable. Seaman (1970, 298) estimates that factory closures as a result of take-overs and unfavourable inter-factory comparisons by national multi-plant firms cost south-east London some 20,000 jobs in the short period 1964–70 alone, representing a decline of over 12 per cent in the area's manufacturing employment. Although ignored by most movement studies, it is clear that such developments may have a considerable effect upon industrial location patterns, are tending to increase in frequency, and in many cases represent a response to forces, including government employment mobility policy, which in smaller firms would lead to movement as traditionally defined. It is worth noting that for this reason alone, Howard's not-inconsiderable totals of post-war employment movement to the peripheral areas probably underestimate the full locational impact of government regional development policy.

Prediction and potential mobility

From a planning viewpoint, the problem of movement definition is however much less important than that of movement prediction. Movement prediction is necessary at two separate levels at least. At the national, aggregate, level, prediction of probable future gross movement rates is of major importance for regional economic planning. Although not investigated here, the demonstrated relationship between inter-industry growth and mobility rates suggests that for manufacturing at least, post-war variations in annual movement rate (Figure 2.3) might well be closely correlated with fluctuations in the national economic growth rate. The intensity of application of government industrial mobility policy, in the form of IDC control and financial inducements to movement, has undoubtedly been an additional important influence. As Figure 2.3 shows, the two periods of maximum post-war manufacturing movement were both associated with unusually vigorous prosecution of government mobility policy (McCrone, 1969, 112, 120). Although exact figures are not yet available, the major new inducements and pressures introduced by the recent Labour government seem to have been associated with a continuing high migration rate since 1965, the values for both 1966 and 1967 being of the order of 200 migrant establishments per year.

However, movement prediction, if only in probability terms, is perhaps of greater importance at the more disaggregated level of individual industries or even firms. As has already been pointed out, some industries and types of firm seem to have been far more mobile than others in post-war Britain. For government mobility policy to be fully effective, therefore, planners ought perhaps to be able to identify, and concentrate their publicity and efforts upon, those firms which are potentially most mobile at any given time (Smith, 1970, 196). While final decision on the granting of individual IDCs must admittedly await the detailed information which only individual IDC applicants can provide, some indication of the types of firm which are most likely to consider movement, and move successfully, cannot but be of value to planners. This may be especially true in relation to ODP policy, which has frequently been accused of a blanket approach to the control of central London office growth, irrespective of the claims of some of the area's office activities to a central location on grounds of close

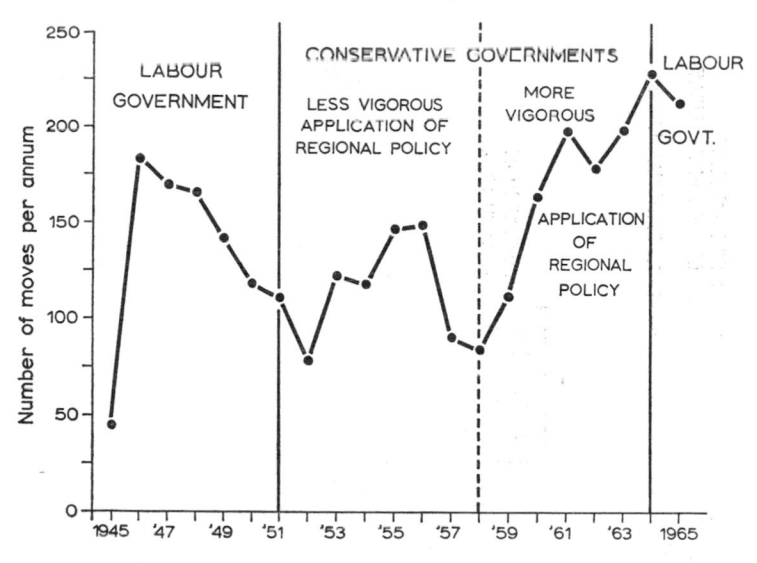

Figure 2.3. United Kingdom: manufacturing movement rates, 1945–65.

Note: The figures relate only to those moves still operating in 1966. Movement rates for the early post war period thus understate the actual amount of movement occurring at that time, since a number of factories then established have subsequently ceased production. The selection of 1958 as the division between periods of more and less vigorous regional policies pursued by Conservative governments follows Cameron and Clark (1966, 18).

Source: Howard, 1968, Appendix A, 39.

functional ties to the area and importance in invisible exporting. A selective approach to mobility promotion must however grapple with at least two major problems.

One is the fact that, as Haig pointed out as long ago as 1926, economic enterprises cannot be regarded as organically unified entities, requiring a single location at which to perform a single clearly-defined function. Rather, 'every business is a packet of functions, and within limits these functions can be separated and located at different places' (Haig, 1926, 416). In other words, firms are themselves aggregates of different activities, which may possess different locational needs and different movement probabilities. Figure 2.4 attempts to portray the main distinctive functions or activities carried on in lesser or greater degree by manufacturing firms. Arrows indicate the direction and degree of locational interdependence of different activities, while examples of the locational influences likely to affect each activity are also noted. As the diagram shows, the four activities most likely to occur in different locations are end-product manufacture, component

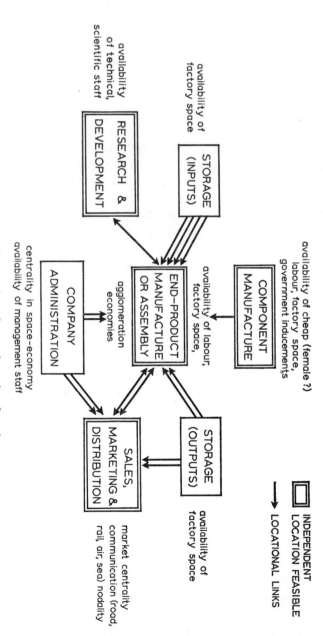

Figure 2.4. The packet of manufacturing functions.

COMPONENT
MANUFACTURE

availability of cheap (female ?)
labour, factory space,
government inducements

STORAGE
(INPUTS)

availability of
factory space

RESEARCH &
DEVELOPMENT

availability
of technical,
scientific staff

END-PRODUCT
MANUFACTURE
OR ASSEMBLY

availability of labour,
factory space,

COMPANY
ADMINISTRATION

agglomeration
economies

centrality in space-economy
availability of management staff

SALES,
MARKETING &
DISTRIBUTION

STORAGE
(OUTPUTS)

availability of
factory space

market centrality
communication (road,
rail, air, sea) nodality

INDEPENDENT
LOCATION FEASIBLE

LOCATIONAL LINKS

manufacture, sales and distribution, and research and development. Though not shown in the diagram, end-product manufacture may itself be split between two or more locations, if the firm manufactures different products or elects to serve geographically separated markets by two or more plants manufacturing the same product. Of course, many firms do carry on all the activities specified in the diagram from a single location. Small firms in particular usually cannot afford the frictional costs of moving goods or people between different establishments. But as firms grow larger, and internal specialisation increases, geographical separation and movement of particular departments becomes more likely. The packet of functions concept is thus of considerable relevance in the analysis and identification of potentially mobile activities, especially where large manufacturing or office firms are concerned.

The second problem is how to establish, even at the firm level, logical criteria by means of which to identify firms with high movement probabilities. The most obvious criterion is perhaps growth rate. As already noted, expansion usually provides the fundamental stimulus to movement through increased pressure on and cost of premises, local labour supplies, etc. However in many ways, knowledge of industry or firm growth rates is much less helpful to planners seeking to implement government mobility policy than may seem to be the case. All else being equal, most growing firms prefer to expand *in situ*. The planner's real problem, especially when concerned with the issue of IDCs or ODPs, is to distinguish between growing firms which are strongly tied to their existing location for sound economic reasons, and those which are not. Clearly, this brings discussion back to the problem of measuring 'potential mobility', as defined earlier.

That ties to a local area may be of major importance even for rapidly expanding firms is certainly suggested by the very limited success of office mobility policy in recent years, despite the imposition of remarkably stringent ODP controls on further office building in central London. Although net office employment growth in that area has virtually ceased, soaring office rents and acute office space shortages have not stimulated increased movement of office firms to other areas. On the contrary, office movement has in fact declined significantly in the last few years, despite the continuing and valiant efforts of LOB. Since 1967–8, the annual movement rate from central London has fallen from 191 firms (14,002 jobs) to only 129 firms (9,367 jobs) in 1969–70 (Location of Offices Bureau, 1970, 5). This decline, which largely reflects the exhaustion of available 'decentralised' office space in outer London, clearly illustrates the very marked reluctance of central London office firms to move more than a few kilometres, if at all, from their existing location. So too does the fact that the total number of decentralised jobs reported by LOB (63,330, 1963–70) represents no more than 8 per cent of central London's 1961 office employment.

Evidence from a variety of independent studies suggests that the very low potential mobility of central London office firms primarily reflects the strong functional linkages binding many of these firms to one another and to focal institutions such as the Stock Exchange or Lloyds (Economist Intelligence Unit, 1964; Dunning, 1969). This finding suggests that for offices of the type found in central London, a possible measure of potential mobility might be some form of local linkage intensity index. The problem here is that measurement of local office information linkages, which may take the form of personal meetings, telephone calls or messenger transfer of documents, is complex and time-consuming. Indeed, Goddard's current study (1970, 53) of variations in local linkage intensity between different London office firms, results of which are not yet available, is the first of its kind ever to be carried out in Britain. Moreover, though probably useful in the office case, linkage intensity measurement seems to be of only limited value as an index of the potential mobility of manufacturing firms. Thus a recent study of local North-West London manufacturing linkages, defined in terms of the interchange of semi-processed goods or components between local firms, concluded (Keeble, 1969, 181) not only that such linkages were of possible locational significance to only between 25 and 30 per cent of firms in the area, but that most of these firms were in the very industry group (engineering and electrical goods) which had contributed most to post-war movement from North-West London. Only in the case of certain limited categories of specialist engineering firms, notably those engaged in subcontract work and metal-processing, was movement markedly inhibited by linkage relationships. Though probably more significant in other industrial areas, such as the clothing quarters of inner London or the West Midlands Conurbation with its metal-fabricating trades (Townroe, 1971), local linkage dependency would thus seem of limited general value as an index of the potential mobility of manufacturing firms.

A possible alternative in the manufacturing case is the strength of local or sub-regional market orientation, for end-products as well as semi-finished goods. As Figure 2.5 shows, migrant North-West London firms tend to be much less dependent upon customers in Greater London, as measured by deliveries, than firms which still operate in the area. Calculation of χ^2 indicates that the difference is highly significant, at the 0.001 level. Moreover, it cannot be explained by market re-orientation following movement on the part of the migrant firms, since the data basically refer to the

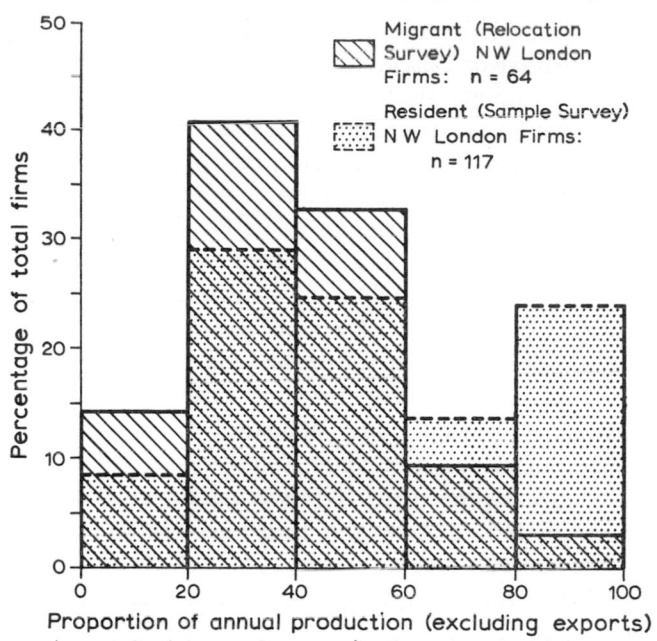

Figure 2.5. United Kingdom: local market orientation and manufacturing movement from North West London.

Source: N.W. London industrial location and relocation survey, unpublished data.

situation at the time of move. The most striking feature of the graph is the difference between the proportions (3 per cent as compared with 24 per cent) despatching 80 per cent or more of production to Greater London customers. Such firms have clearly played very little part indeed in movement. Conversely, firms which are less dependent upon London customers have been more mobile than might be expected from their relative frequency in the area. This evidence does suggest that local market orientation may be a useful guide to the potential mobility of manufacturing firms, especially at the extreme ends of the possible range of dependency values.

The evaluation of mobility patterns

Discussion of possible diagnostic indices of movement probability leads naturally to consideration of the final, major, question posed earlier – what factors explain variations in movement streams, notably of manufacturing and office firms, between different parts of the country, and how best may these factors be identified and their relative importance evaluated? A concern for explanation of observed spatial variations in movement lies behind most of the mobility studies carried out in recent years. Its most recent, and perhaps most striking, expression is the major survey of industrial movement between 1964 and 1967 recently carried out by what is now the Department of Trade and Industry's Regional Economics and Statistics Division under the direction of its Principal Research Officer, R. S. Howard. This survey, known as 'the Inquiry into Location Attitudes and Experience' (ILAG), has sought to investigate in detail the experience, reasons for movement, method of search and locational choice of all manufacturing establishments set up in the United Kingdom during the period 1964–7 inclusive as a result of movement across one of the boundaries defining the Board of Trade's fifty statistical areas. Detailed replies, mostly from direct interviews, to a comprehensive questionnaire have been obtained from no less than 630 firms, representing the great majority of the 800 or more moves recorded by the Board of Trade for this period. Publication of the survey's results, due in 1971, will clearly add greatly to current knowledge of the reasons for spatial variations in manufacturing movement in Britain.

However, though undoubtedly the most ambitious and far-ranging survey of actual industrial location decisions ever undertaken in Britain, the ILAG inquiry is not conceptually different from the great majority of mobility studies carried out in recent years in Britain and North America. In adopting an analytic framework which seeks to evaluate locational choice by asking direct attitude-testing questions of firms' representatives, however large the sample of establishments studied, the inquiry conforms to what may be termed the 'micro-level' approach of most recent studies. Such studies, of which those by Cameron and Clark (1966), Wabe (1966), Loasby (1967), the present author (Keeble, 1968) and Townroe (1971) provide good examples, are usually based on the results of first-hand surveys of samples of individual migrant firms. While necessitated until recently by the complete absence of published official statistics, this survey approach also reflects the implicit or explicit view that the most logical method of explaining movement and the choice of new locations is to ask the decision-makers actually involved. Micro-level studies thus almost invariably deal only with samples of mobile firms, rather than total populations, and often present their findings on locational choice in the form of tabular listings of the frequency with which particular location factors were reported as influencing the firms investigated.

Micro-level analysis has been of the greatest importance in enhancing understanding of the geography of manufacturing and office movement in Britain. It has yielded invaluable and otherwise unavailable information on a wide variety of topics, ranging from the economic viability of factories set up in the relatively isolated peripheral regions (Luttrell, 1962), to the journey-to-work implications of office decentralisation within the London city-region (Daniels, 1970). However, it is known to suffer from certain defects. One is that a sample-based approach may lead to inaccurate findings because of under- or over-representation of particular types of migrant firms, relative to their importance in the whole migrant population. This is especially a problem in mail-questionnaire surveys, which may yield response rates of as little as 40 per cent (Hunker and Wright, 1963, 11). Another is that questions on locational influences sometimes result in confused answers because of a failure to distinguish clearly between factors operating only at an inter-regional scale (such as government financial inducements), and factors which are really significant only at an intra-regional or local level (such as the availability of a modern factory, or effluent disposal facilities). A third is the problem of *post-facto* rationalisation, which leads firms to ascribe their decisions to locational advantages which in fact only became apparent after movement. A similar situation is the cloaking of more personal, perhaps economically-irrational, motives by reference to 'acceptable' considerations of a business character which in fact did not really influence the decision at all. This sort of inaccuracy is particularly common in questionnaire replies by smaller firms. On the other hand, such replies are at least much more likely to be provided by the actual decision-taker himself, in the form of the managing director (and often founder) of the firm; whereas the junior management staff who are often given the task of questionnaire response in large firms may well be ignorant of the real factors behind a board-room movement decision.

These various problems suggest a need for an alternative, and if possible complementary, approach to micro-level analysis of employment mobility. Such an approach is common in studies of the movement of other human phenomena, such as people, traffic or goods. Such studies are more often than not based on statistical manipulation of aggregate data, in which spatial variation in total movement, or in movement estimates based on large samples, is investigated. In such 'macro-level' analysis, direct behavioural questioning is replaced by operational definition and statistical testing of hypothesised relationships between variables presumed to influence movement variations, and variations in movement itself. The hypotheses tested may be a product of theoretical, often deductive, analysis: or they may be rooted in earlier empirical findings, perhaps of micro-level studies. Quite often, the hypotheses are expressed in the form of mathematical models, whose outputs are compared with the observed reality of movement patterns. The most common of these are interaction models, based originally on the concepts of social physics and the gravity analogy. More recent model-building approaches in macro-level movement studies include Markov-chain analysis and the use of Monte Carlo simulation models.

The value of macro-level analysis in movement studies of phenomena other than business firms is now well established. Why then has it hitherto been almost totally neglected by geographers and location analysts concerned with employment mobility? The reason of course is that it depends on the availability of aggregate, representative or comprehensive statistics, defining with reasonable accuracy real world movement patterns. Such information, for manufacturing movement in the United Kingdom, has only recently become available, with the publication of the Board of Trade data (Howard, 1968). No comparable source exists for statistics of office movement, although LOB's records, which are inevitably incomplete, might be used in this way.

Publication of the Board of Trade data thus for the first time makes possible an attempt to explain spatial variation in manufacturing movement by macro-level analysis. Moreover, earlier micro-level studies provide an obvious source of hypotheses which can and need to be tested at the aggregate level. Having surveyed the main findings of previous micro-level analysis, an attempt will therefore be made to provide one or two very simple illustrations of the way in which such testing can be carried out.

Micro-level studies: findings

One of the more important findings of micro-level studies of employment mobility concerns what may be termed the dual-population hypothesis of industrial movement. This hypothesis suggests that if local intra-urban movement is ignored, the great majority of individual moves in post-war Britain can be classified into one or other of two distinctive categories. Though fundamentally differentiated by distance of movement and nature of reception area, these categories also yield populations of migrant establishments which differ significantly in terms of establishment size and organisational type. In short, the dual-population hypothesis states that post-war movement of manufacturing, and to some extent office, activity in Britain may basically be divided into long-distance, large-establishment movement to peripheral regions of higher than average unemployment, and short-distance, small-establishment urban overspill movement around major conurbations, particularly in the more prosperous 'central' regions of the country. In the manufacturing case, the former inter-regional category is dominated by the setting up of branch factories, complete transfers of existing firms being much less common. Many of the establishments in the latter intra-regional category, however, have been set up by firms which have completely transferred their activities from the central conurbation of the region.

Micro-level evidence for this hypothesis, at least in terms of manufacturing activity, is considerable. Perhaps most important are those studies (Loasby, 1967; Keeble, 1968) which have investigated the character and spatial pattern of movement from one particular origin area to all other parts of the country. Both these analyses stress that the basic locational choice made by firms from conurbations such as London and Birmingham has been between one of the peripheral regions of higher unemployment, in which government incentives were available, and the area surrounding the conurbation concerned. Both draw attention to the size difference of establishments set up in these two alternative types of location. Both point to a significant variation in the balance between branch plants and transferred establishments in the two types of movement. And both suggest, either implicitly or explicitly, that these contrasts are linked to differences in the perceived locational attractions of the two sorts of reception area. Moreover, their findings are supported by other studies of particular reception areas. Thus Brown's analysis (1965) draws attention to the small size of many migrant establishments settling in the London new towns (nearly half – 44 per cent – employed less than 50 workers in 1963); Twyman (1967) has shown that movement to east Kent, primarily from London, has involved a majority of relocated plants (66 per cent); and Law (1964) has demonstrated that industrial migration to Northern Ireland has been dominated by the establishment of self-contained branch factories.

The chief perceived locational attractions to manufacturing industry of the peripheral regions, as revealed by firms' replies to these micro-level surveys, are three-fold. Most important is labour availability, evident in higher-than-average unemployment rates, lower-than-average activity rates, and net out-migration, chiefly of younger people of working age. All micro-level surveys dealing with the peripheral regions reiterate the importance of this factor to firms from the labour-shortage areas of south-east England and the Midlands. A typical finding is that of Cameron and Clark (1966, 164), who stress that as far as their sample of 71 assisted area moves were concerned, 'the supply of trainable labour was the most important single determinant of area/site choice'. Labour costs have not been viewed as significantly lower in these areas, and have not generally entered into the locational equation. The second major influence, of equal importance with labour availability, has been government pressure and inducements. The various forms which government action has effectively taken include refusal of IDCs for new premises anywhere other than in a peripheral region, financial inducements to firms locating there (especially significant in the nineteen-sixties), and the provision, often ahead of demand, of available factories on government-sponsored estates. Cameron and Clark's conclusion (1966, 67) that 'if the Government had not operated a policy of control and inducement, the amount of freely-chosen movement to the assisted areas would have been extremely small' again summarises the findings of virtually all micro-level studies dealing with these locations. The remaining locational influence, of much less importance to the overall movement pattern, has been the presence of fairly large regional markets in certain peripheral regions which have attracted market-orientated branch plants (Keeble, 1968, 33–4; Cameron and Clark, 1966, 80–2).

Micro-level manufacturing movement studies thus suggest that selection of a peripheral, rather than more central, region of the country may be explained by only three main locational factors. Within the group of peripheral regions, however, selection of a particular location has also been affected by the further variable of distance from the firm's origin. Distance minimisation is a surrogate for various important influences upon locational choice, including ease and speed of search, the need to maintain contact with the parent factory, existing suppliers and customers, and feasibility of persuading key workers to settle in the new location. Its influence is well illustrated by Loasby's finding (1967, 37) that Birmingham firms deciding on a peripheral region 'usually began by examining the area nearest to their original location. For this reason, South Wales was clearly the favourite among development districts...and within South Wales, there was a clear preference for most easterly sites'. Cameron and Clark (1966, 201) reach a similar conclusion in studying moves to the whole range of peripheral regions.

As pointed out earlier, the dual-population hypothesis implies that short-distance, intra-regional overspill from the major conurbations of the country has been influenced by locational considerations significantly different from those affecting peripheral movement. Again, this is largely substantiated by micro-level surveys. The fundamental influence upon locational choice reported by firms moving out of London and Birmingham to surrounding settlements is a perceived need for proximity to the particular conurbation concerned. Thus Brown (1965, 126) concludes that 'the most binding requirement' of firms moving to the London new towns 'seems to be the need for proximity to London', while Loasby (1967, 37) reports an exactly similar finding for movement from Birmingham to surrounding planned overspill centres. As with the distance factor in peripheral movement, the proximity variable (which accounts for the rejection of all peripheral regions by these short-distance moves) represents a surrogate for a variety of influences (Keeble, 1968, 37–41). Proximity to existing customers, suppliers and export outlets, access to parent factories with whose production a branch plant's activity may be closely integrated, the intangible but perceived importance of nearness to the centre of technological change in a particular London- or Birmingham-based industry, and social preferences which may have economic expression in the ease or difficulty of retaining workers or attracting new staff, all seem to play a part.

At the 'between-region' level, rejection of peripherality, despite the locational attractions of the peripheral regions, is thus nearly always explained by a desire for proximity to the original location of an intraregional move. However, this factor apparently also operates at the 'within-region' level, to influence choice of reception settlement within, say, south-east England. This point, suggested by Loasby, is clearly illustrated by the North-West London study's finding (Keeble, 1968, 26) of a surprisingly regular gradient of migrant factory frequencies with distance from the origin area up to about 160 kilometres (100 miles) away. The regularity is surprising because of the undoubted impact of other influences upon locational choice within a given central region. Four of these other influences can be singled out as significant constraints on movement patterns at this more detailed geographical scale.

The most interesting is the marked tendency, first investigated in London and south-east England (Keeble, 1965, 27–8) but recently strongly corroborated for a very large sample (*c.* 700) of mobile Birmingham firms (B. M. D. Smith, personal communication; c.f. Townroe, 1971), for radial manufacturing movement. In other words, firms more often than not choose to move outwards from London or Birmingham to settlements in the same compass sector as that in which their original conurbation factory was located (Lewis, 1971, 204). This seems to be associated with the dominantly radial communications pattern in these regions, especially the south-east, and may well be another spatial expression of the conurbation proximity factor. A further, perhaps surprising influence, is the attraction of labour-hungry firms to local areas of higher-than-average unemployment. South-east England in particular possesses a number of peripheral settlements, many of them resorts, where percentage unemployment is unusually high. Indeed, averaged over the three years 1958–60, the absolute total of unemployed workers in the South East standard region outside Greater London was no less than 30,000. As Twyman (1967, 18–19) has demonstrated for east Kent, this labour availability factor has modified radial movement patterns in the south east by attracting to south coast towns a significant number of firms originating in London north of the Thames.

Reference to local areas of higher-than-average unemployment introduces the question of planning controls and influences. In the central regions, IDC and factory availability have played an important role in channelling much manufacturing movement either to local unemployment areas such as east Kent and Portsmouth, or to planned overspill settlements such as the new and 'expanded' towns of the west Midlands. In the case of overspill settlements, the possibility of wholesale transfer of an existing workforce has been a major additional attraction to some industrialists. However, at least in the case of south-east England, new and expanded town movement is in some ways only an aspect of radial migration, since not only has the designation of such towns been influenced by the anticipated possibilities of radial migration in each sector of the region, but organisation of movement to particular new towns was originally deliberately conducted on sectoral lines. The remaining influence worth mentioning is the attraction of desirable residential areas to small firms whose entrepreneurs are personally involved in the consequences of their locational decision, or to concerns employing unusual numbers of more highly-qualified research staff.

Micro-level studies of office movement suggest that the dual-population hypothesis has considerable significance in this context also. A major point here, of course, is that office movement on any scale in Britain has effectively been confined to decentralisation from London, especially central London. The unique importance of the latter small area as the country's greatest office centre is clearly illustrated by the fact that in 1963, the rateable value of offices in inner London (the former LCC area) represented no less than three-quarters of the total value of all offices in England and Wales (Hammond, 1964, 132). Perhaps the most striking finding of micro-level office mobility studies is that movement from central London by commercial firms has been very largely restricted to short-distance migration on the intra-regional scale. Thus Wabe (1966) points out that only 13 per cent of all central London moves (114) recorded by LOB as undertaken or planned between 1963 and the end of January 1965 involved distances of 64 kilometres (40 miles) or more; while Daniels (1969) reports a figure of only 11 per cent for moves by private central-London firms over distances greater than 140 kilometres (90 miles). These studies clearly suggest that London office firms are far more reluctant to participate in long-distance migration than are metropolitan manufacturing plants, the great bulk of office movement being of the overspill type to suburban London and surrounding settlements. Not surprisingly, 'proximity to central London' seems to be regarded by most decentralising office firms as a binding requirement restricting their choice of a new location to the south east of England. This is almost certainly connected with the very high proportion of small office moves, involving less than 60 employees, in intra-regional movement. Daniels' maps (1969) suggest that these make up no less than 70 per cent of all moves from central London to other parts of the South East Planning Region. As a result, the average number of jobs per move involved in this intra-regional movement is only 78, according to Daniels' figures. And this estimate, based on partial information, is almost certainly on the high side.

At the same time, micro-level studies also indicate that what inter-regional office movement has occurred also conforms in many ways to the dual-population hypothesis. Such movement differs markedly from that over shorter distances in being dominated by the transfer of government offices to locations in the peripheral regions. Major examples of this unusual, and undoubtedly politically inspired, movement include (Hammond, 1967) the transfer to Durham of the Post Office's Savings Certificate Division and Savings Headquarters (2,500 jobs), to Glasgow of the Post Office Savings Bank (7,500 jobs), and to Chesterfield of the Post Office's Accountant General's Department (1,700 jobs). Daniels (1969) suggests that in fact no less than 80 per cent of all office jobs involved in inter-regional movement from central London are in government establishments. The large number of jobs involved in several of these government office moves means that on Daniels' figures, the average size of all migrant establishments involved in inter-regional movement is 290 employees, or nearly four times larger than the corresponding figure for intra-regional movement. However, as Wabe (1966) points out, this apparent relationship between size and distance of move does not hold nearly so strongly for private offices considered alone, since many large private central-London offices are apparently as reluctant as small firms to move further than suburban London locations.

Macro-level analysis and manufacturing movement

As already pointed out, macro-level testing of the findings of micro-level studies must perforce be confined to manufacturing movement, for which aggregate Board of Trade data are available. At the simple level of

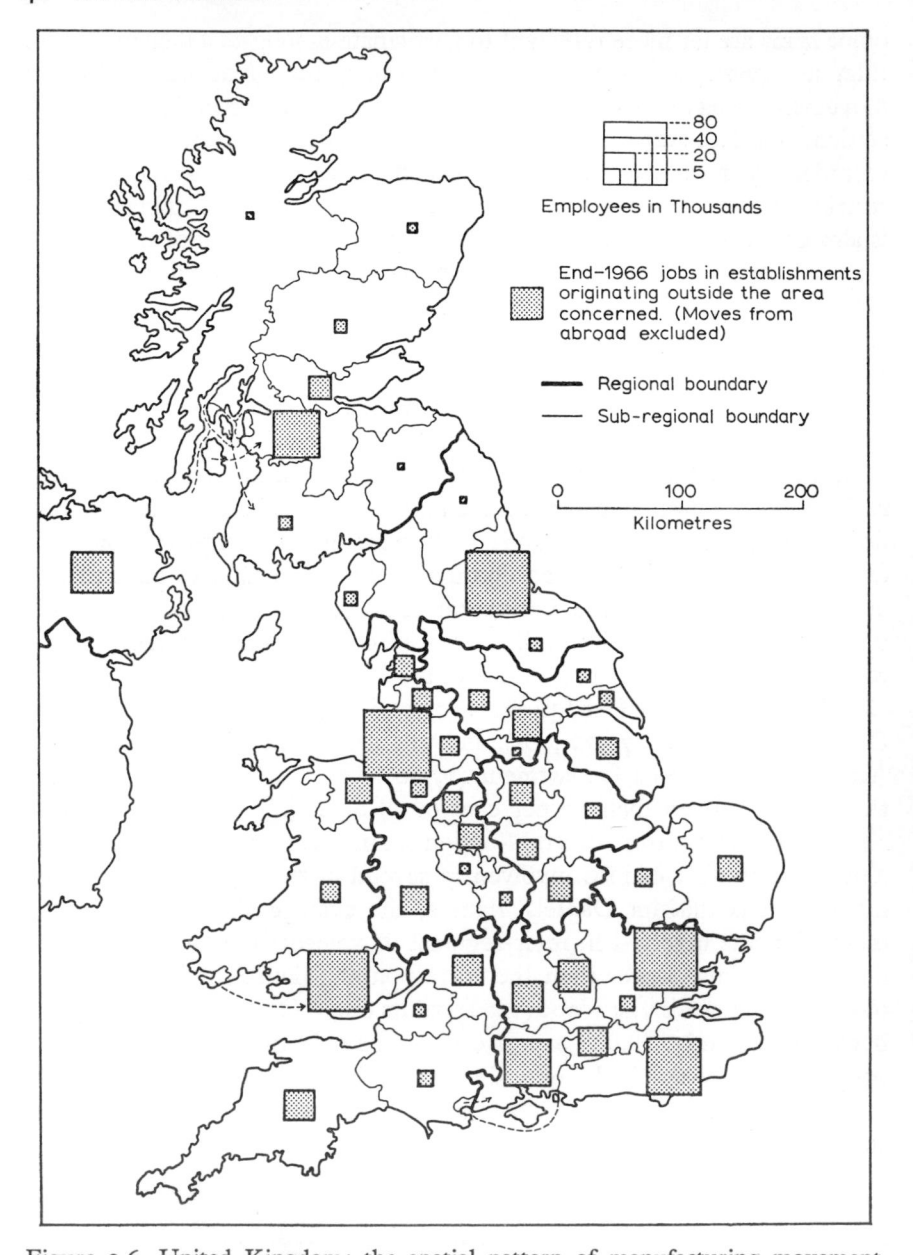

Employees in Thousands

End–1966 jobs in establishments originating outside the area concerned. (Moves from abroad excluded)

—— Regional boundary
— Sub-regional boundary

Figure 2.6. United Kingdom: the spatial pattern of manufacturing movement, 1945–65.
Source: Board of Trade 50 × 50 movement matrix.

descriptive statistics, Howard's study (1968) provides surprisingly strong support for the dual-population hypothesis. Its data, and that contained in a slightly fuller version of the 50 by 50 movement matrix kindly made available by the Board of Trade, reveal for example that no less than 72 per cent of all moves (as defined in the study) within the United Kingdom between 1945 and 1965, and 77 per cent of all jobs thus created by 1966, fall into one or other of the categories suggested by the hypothesis. These calculations exclude 'movement' by foreign firms setting up for the first time in the United Kingdom, as well as a small number of moves which could not be allocated by the Board to a region of origin. Inter-regional movement to the six peripheral areas distinguished in the study involved 921 establishments (345,000 jobs), while intra-regional movement from those central conurbations or cities (Greater London, the West Midlands, Leeds and Glasgow/Edinburgh) for which approximate data are available involved 1,038 establishments (232,000 jobs). The dominance of these two types of movement stream is suggested by Figure 2.6, which maps gross aggregate totals of jobs created by migrant manufacturing establishments in each of the 50 areas distinguished by the Board of Trade. Jobs provided by foreign firms beginning production for the first time in the United Kingdom are again excluded. The map confirms in visual terms the importance of movement to south-east England outside Greater London, and to the peripheral regions (notably South Wales, Merseyside, north-east England, central Scotland and Northern Ireland). The movement matrix shows that the greater part of the former is London overspill, while over 95 per cent of the latter, measured by jobs, originated in the central regions of the country. Movement between or within particular peripheral regions has thus been of negligible importance, while flows from the peripheral regions towards the centre are virtually non-existent, only 20 cases (out of 2,728) being recorded of movement from any of the peripheral areas to either the South East or West Midlands regions. The detailed pattern of moves from the two dominant movement origins, Greater London and the West Midlands Conurbation (which together generated no less than 56 per cent and 53 per cent of all United Kingdom moves and migrant jobs, respectively) is shown in Figure 2.7. In both cases, the spatial distinctiveness and dominance of the two types of movement stream postulated by the dual-population hypothesis is strikingly apparent.

Moreover, the Board of Trade data also support this hypothesis as it concerns factory size and organisational character. Although not discussed in detail by Howard (1968), average migrant establishment size can be shown to differ significantly between the two types of destination area. The

4

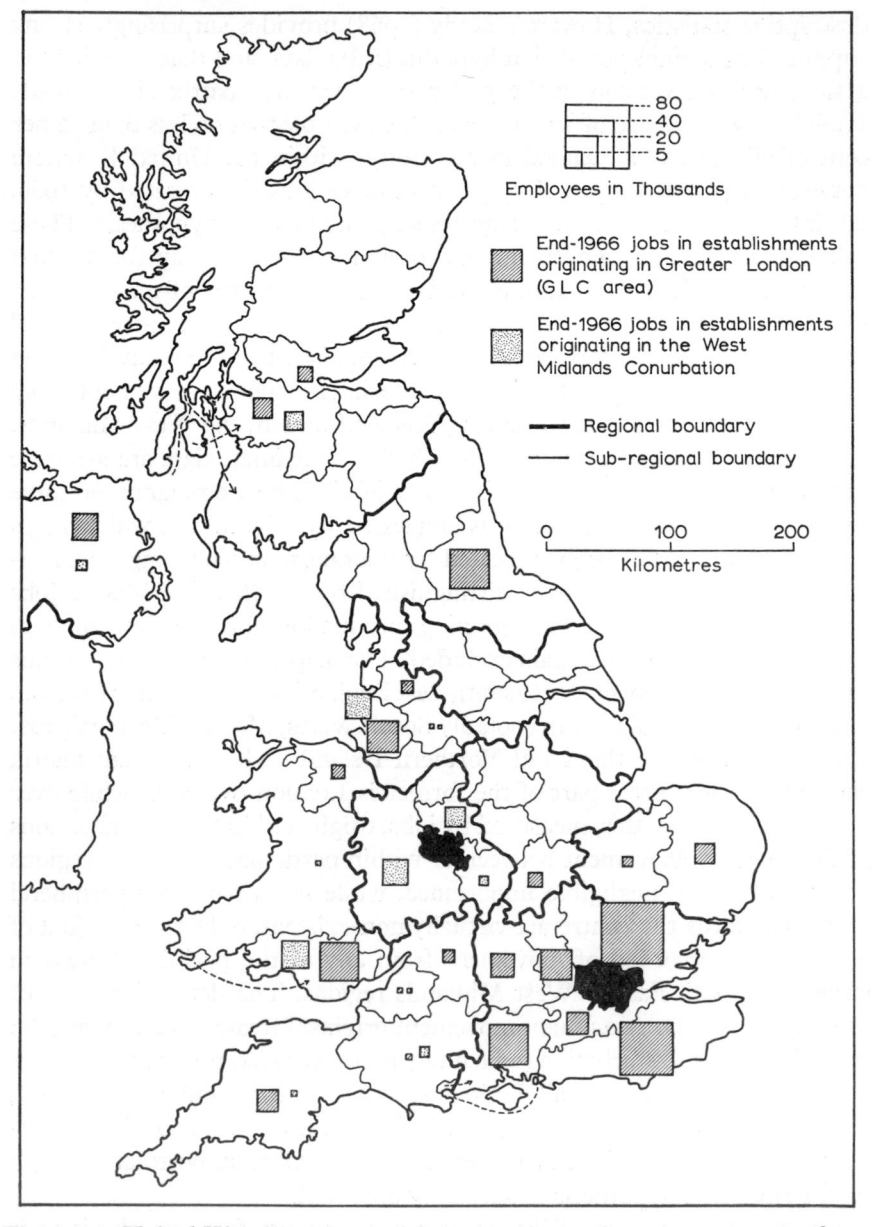

Figure 2.7. United Kingdom: the spatial pattern of manufacturing movement from the Greater London and West Midlands conurbations, 1945–65.
Source: Board of Trade 50 × 50 movement matrix.

TABLE 2.1. *United Kingdom: mean factory size and movement to central and peripheral areas*

Central areas	Mean migrant factory size (employees)	Peripheral areas	Mean migrant factory size (employees)
Hampshire	345	Merseyside	706
Berkshire and Oxfordshire	321	Northern	414
Surrey and west Kent	302	Wales	324
Sussex and east Kent	251	Northern Ireland	317
Essex, Hertfordshire and Bedfordshire	228	Devon and Cornwall	269
Staffordshire	205	Scotland	247
Shropshire, Herefordshire, Worcestershire and Warwickshire	196		
Buckinghamshire	194		
East Anglia	137		
Central areas' mean	252	Peripheral areas' mean	367

Sources: Howard (1968) and Board of Trade movement matrix.

scatter diagram of Figure 2.8 and Table 2.1, based on calculations from the full movement matrix, reveal that while mean factory size varies considerably within each of the two sets of peripheral and central overspill areas, the overall mean values for each set are significantly different. That for migrant factories in the six peripheral areas is 367 employees, compared with only 252 employees for factories in the nine central overspill areas plotted. The latter represent all the main reception areas for short-distance industrial overspill from London and the West Midlands. As the diagram and table show, these set mean values are in fact respectively greater and smaller than any individual area value in the opposite set, with the single exception of Scotland. Interestingly enough, the movement total for this peripheral region is the only one to include any significant volume of intra-regional migration (48 cases, involving 8,5000 jobs). If these are omitted from the calculation, as would seem logical in the context of testing the dual-population hypothesis, the Scottish value rises to 265 employees, above the central area set mean.

Although affording less detailed information than that relating to factory size, the Board of Trade data also support the dual-population hypothesis of differing tendencies in factory organisation as between inter- and intra-regional movement. Howard's study (1968, 23) reveals that movement to

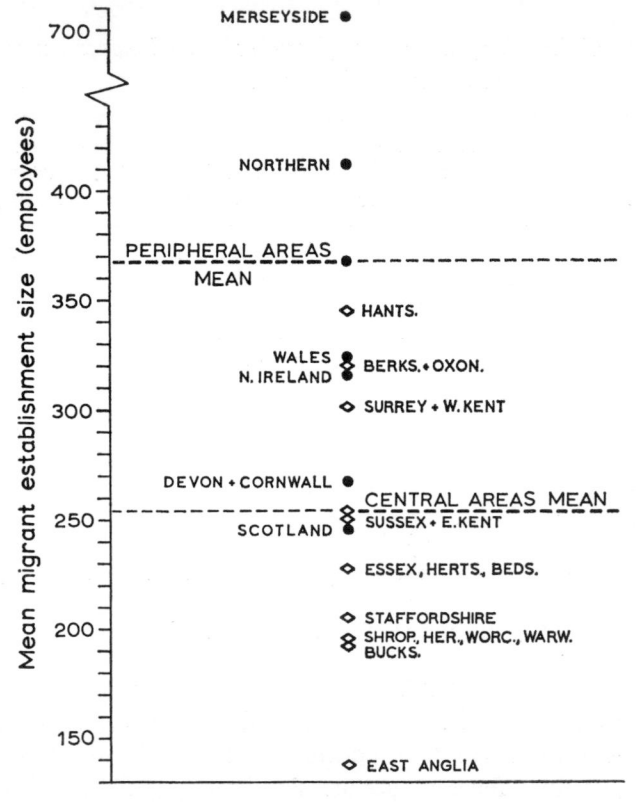

Figure 2.8. United Kingdom: factory size and the dual population hypothesis.

● Peripheral Areas   ◇ Central Areas

Note: Areas for which movement totals of less than 10,000 workers were recorded have been amalgamated with adjacent areas so as to eliminate freak values resulting from small samples. Inclusion of East Anglia, officially a separate economic region, is fully justified in view of its nearness to London and the almost complete dominance of movement to it by London firms of the typical overspill variety (Keeble, 1968).

Source: Board of Trade 50 × 50 movement matrix.

the peripheral areas has indeed been dominated by the establishment of branch factories. The latter accounted for 83 per cent of all peripheral area moves, 1945–65, and 86 per cent of all peripheral area jobs thus created by 1966. However, movement to the South East and East Anglia, largely reflecting outward decentralisation from London, has in contrast been characterised by a higher proportion of complete transfers (57 per cent) than of branches, although the small size of many of these transfers means that branches still account for the greater share (56 per cent) of employment created.

Descriptive statistics culled from the Board of Trade movement matrix thus provide fairly strong support for a dual-population hypothesis defining two distinctive types of manufacturing movement streams in post-war Britain. However, this hypothesis and micro-level findings also suggest the operation of different sets of locational influences upon the two types of movement. How may the Board of Trade statistics be used analytically at a macro-level to investigate this more complex issue? Two very simple illustrations, concerning respectively long-distance movement to the peripheral regions, and short-distance urban overspill in central areas of the country, will be presented.

One of the more interesting aspects of long-distance manufacturing movement to the periphery, as revealed by the Board of Trade data, is the considerable variation in movement totals (measured by jobs created) between different peripheral regions. As Figure 2.6 indicates, gross aggregates vary from 87,000 jobs (Merseyside) to only 18,000 jobs (Devon and Cornwall). The range of peripheral totals for movement from particular major origin areas, such as Greater London and the West Midlands Conurbation, is relatively even greater (Figure 2.7 and Table 2.2). As already pointed out, micro-level studies suggest that movement to particular peripheral regions has been influenced by four main factors – labour availability, government influence and inducements, access to markets (especially regional markets), and distance from a given area of origin. Of these, government influence may be regarded as not discriminating spatially between the six peripheral regions considered, all of which are Development Areas and have benefitted from government action for a reasonably long period of time. Micro-level findings thus imply that spatial variations in movement to different peripheral regions should be explainable in terms of variations in labour availability, in distance from major origin areas, and to a lesser extent in market access, especially to regional customers.

Consideration of these factors in the context of macro-level evaluation of the reasons for spatial variation in movement immediately suggested the use of some sort of gravity-model formulation. In a gravity model, the positive attractiveness of a given destination area, quantified in some way, is discounted by some measure of the distance over which movement has to take place. In this context, use of such a model implies that industrialists selecting a new location from amongst the set of peripheral regions on offer consider simultaneously the two variables of intrinsic regional attractiveness and distance of move. In other words, a region which is inherently more attractive but further away may exert the same overall pull on industry from a given area as one which is inherently less attractive but nearer. Stress

TABLE 2.2. *United Kingdom: gravity model data and movement to the peripheral regions, 1945–65*

Peripheral regions	Average total unemployed workforce, 1954 (thous.)	Average economic potential	Movement from London (G.L.C. area)		Movement from South East (inc. G.L.C.)		Movement from West Midlands Conurbation	
			Jobs created (thous.)	Airline distance (miles)	Jobs created (thous.)	Airline distance (miles)	Jobs created (thous.)	Airline distance (miles)
N. Ireland	33.0	800	13	320	16.9	320	2	230
Scotland	59.5	1000	13	360	24.5	360	6	250
Northern	28.3	1100	31	250	35.8	250	1	170
Wales	22.9	1100	30	140	43.2	140	16	80
Merseyside	18.9	1300	20	180	36.0	180	12	80
Devon and Cornwall	8.6	900	9	190	10.6	190	1	170

Sources: movement data from Howard (1968) and Board of Trade movement matrix; unemployment data from *Abstract of Regional Statistics* and regional economic planning reports; economic potential data from Clark (1966).

on this gravity model implication of simultaneous evaluation is important if only because the only recent study to utilise the Board of Trade data in a way similar to that illustrated here (Beacham and Osborn, 1970) fails to appreciate this point. Its entirely separate analyses of the relationship between movement and distance, and movement and unemployment rates, thus yield wholly inconclusive results, unlike those recorded below.

The two simple gravity models which were thus constructed as a framework for testing the importance of the variables listed above took the following form:

$$M_{ij} = \frac{A_j}{d_{ij}^b}$$

where M_{ij} is an index of the predicted volume of industrial movement between origin area i and destination area j, A_j is a measure of the intrinsic attractiveness to mobile industry of destination area j, and d_{ij} is the distance between the two areas, raised to some exponent b. For simplicity, it was decided to concentrate attention on peripheral manufacturing movement from only three origin areas, the South East standard region as a whole, the GLC area, and the West Midlands Conurbation, and to analyse movement from each area separately. Separate analysis eliminated the need for calculation of some sort of mass value for each origin area, to allow for variations in outward movement from different origins; while selection of these particular areas is justified by their dominance as origins of inter-regional migration to the periphery, firms from the South East and West Midlands Conurbation having provided no less than 205,000 jobs, or 60 per cent of all peripheral employment resulting from moves from the central areas of Britain. Indeed, no other well-defined origin area for which the Board of Trade provides data contributed as many as 20,000 migrant peripheral jobs which, though arbitrary, would seem a reasonable gross movement threshold for macro-analysis in this context.

In both the gravity models tested, d_{ij} was defined for simplicity as air-line distance measured between the approximate manufacturing centres of gravity of the relevant origin and destination areas. These centres were selected with the help of a map of industrial employment in the United Kingdom. Given the crudity of this distance measurement, which takes no account for example of probable distortions in real-world transfer costs with trans-shipment across the Irish Sea, or with tapered freight rate structures, it was decided to round distance values up or down to the nearest ten-mile unit (see Table 2.2). Thus the difference between the two models lay only in the measurement of peripheral attractiveness, A_j. In view of micro-level evidence, the first and basic model (A_1/d) measured attractiveness solely in terms of labour availability, expressed as the total unemployed labour-force in a particular peripheral area for a year (1954) roughly midway during the period (1945–65) to which the Board of Trade data relate. The use of absolute totals, rather than percentage rates as in the inconclusive Beacham and Osborn study (1970), would seem essential, bearing in mind intended comparison with absolute totals of migrant jobs, and the fact that micro-level evidence clearly suggests that labour-hungry firms considering movement are influenced not by percentage unemployment rates, which can be very misleading indices of labour availability, but by actual numbers of unemployed workers in particular areas.

The second model (A_2/d), while utilising the same basic labour availability measure as the first, attempted also to make some allowance for variations in market accessibility as between different peripheral regions, bearing in mind micro-level evidence on the subsidiary importance of regional market orientation in long-distance movement. The method employed to do this involved weighting of a given region's unemployed workforce total by its average 'economic potential' index (see Table 2.2). Economic potential values were derived from Clark (1966), and represent a measure of the potential, or relative nearness, of any given location with

TABLE 2.3. *United Kingdom: Spearman's rank correlation coefficients*
(r_s) for predicted migration indices against actual migrant
employment in the peripheral areas, 1945–65

Origin area		Distance exponent (b)			
		$d^{1.0}$	$d^{1.5}$	$d^{2.0}$	$d^{2.5}$
G.L.C. area	A_1 model	0.500	0.529	0.614	0.814
	A_2 model	0.500	0.614	0.814	0.729
	Attractiveness (A_1) only: r_s = 0.129. Distance (d^1) only: $r_s = -0.329$.				
South East region (inc. G.L.C. area)	A_1 model	0.772	0.800	0.943**	1.000**
	A_2 model	0.829*	0.943**	1.000**	0.943**
	Attractiveness (A_1) only: $r_s = -0.029$. Distance (d^1) only: $r_s = -0.600$.				
West Midlands conurbation	A_1 model	0.843*	0.929*	0.757	
	A_2 model	0.900*	0.929*	0.757	
	Attractiveness (A_1) only: r_s = 0.072. Distance (d^1) only: $r_s = -0.386$.				

Note: * significant at the 0.05 level ($\geqslant$ 0.829).
   ** significant at the 0.01 level ($\geqslant$ 0.943).

respect to the whole national market for manufactured goods in Britain and to major export outlets. Clark's calculations utilised Inland Revenue personal income data as an index of market distribution, and allowed for real-world tapering of freight rates with distance. Though reflecting the distribution of purchasing power throughout the country as a whole, individual economic potential values are in fact particularly strongly influenced by the scale of nearby, regional market, demand, owing to the weighting procedure adopted. They were thus employed as a readily-available and reasonably-meaningful measure of a given peripheral region's marketing attractiveness to manufacturers, especially in terms of regional sales opportunities. Their use in weighting unemployed workforce totals involved a crude, rule-of-thumb, division of each potential value (Table 2.2) by 1,000, this figure in turn being multiplied by the respective unemployed workforce total to give the A_2 value.

In the gravity model literature, considerable attention is devoted to the general problem of selecting the distance exponent b for a particular model. Fundamentally, this problem arises from the lack of any theoretical or logical grounds (other than the direct gravity analogy of an exponent of 2)

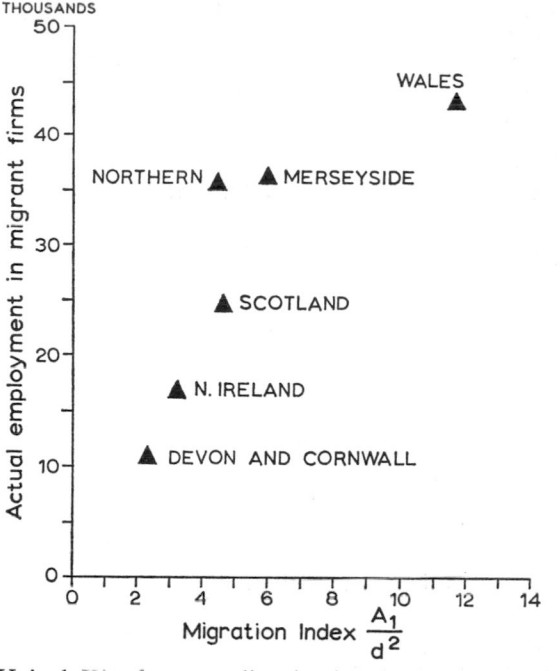

THOUSANDS

Figure 2.9. United Kingdom: predicted migration indices and actual migrant employment in the peripheral areas: movement from South East England, 1945–65. Sources: see Table 2.2.

for specifying what this exponent ought to be in a given case. For this reason, the two models were run several times for movement from each origin area using different distance exponents, including 1.0, 1.5, 2.0 and 2.5. The output of both models took the form of a predicted migration index for each of the peripheral regions. The absolute values of this index cannot be compared for movement from different origin areas. But for movement from a single origin, they do represent a measure of predicted movement to each region which ought to vary proportionately with the actual observed regional movement totals. Because of the small size of sample investigated (six peripheral areas), it was felt that comparison of migration indexes and actual migrant employment by standard parametric correlation methods could well be statistically invalid. It was therefore decided to employ less powerful but more robust non-parametric rank correlation methods (Spearman's r_s), which do not depend upon such stringent assumptions as does the Pearson product-moment correlation technique.

The results of the analysis are given in Table 2.3, and illustrated graphically in Figures 2.9 to 2.12. As the r_s values in brackets indicate, initial

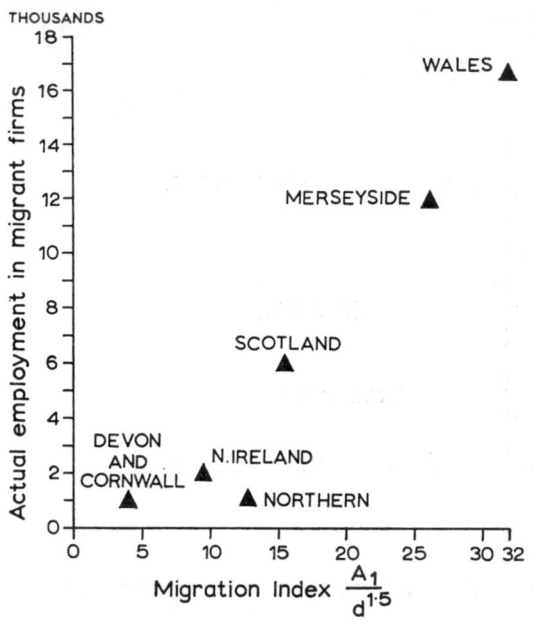

Figure 2.10. United Kingdom: predicted migration indices and actual migrant employment in the peripheral areas: movement from the West Midlands Conurbation, 1945–65.

Sources: see Table 2.2.

correlation of actual migrant employment in each peripheral region with attractiveness and distance values, considered separately, yielded totally insignificant coefficients. However, simultaneous evaluation of these two variables within the gravity-model framework resulted in high and significant correlation values. For movement from the South East as a whole, r_s values of 0.943 and 1.000 (both significant at the 0.01 level) were achieved for the A_1 model with distance exponents of 2.0 and 2.5, respectively (see Figure 2.9). For migration from the West Midlands Conurbation, a distance exponent of 1.5 (see Figure 2.10) yielded an r_s coefficient of 0.929 (significant at the 0.05 level). Only with movement from Greater London did r_s values fail to achieve a 0.05 significance level, although an r_s value of 0.814 was calculated with a distance exponent of 2.5 (higher exponent values resulted in declining coefficients). All these results relate to the A_1 model. Not surprisingly, given the success of this basic approach, the A_2 model failed to achieve higher maximum coefficients. However, as Table 2.3 indicates, weighting by market accessibility did in fact yield correlation values higher than or equal to the A_1 results for a given distance exponent

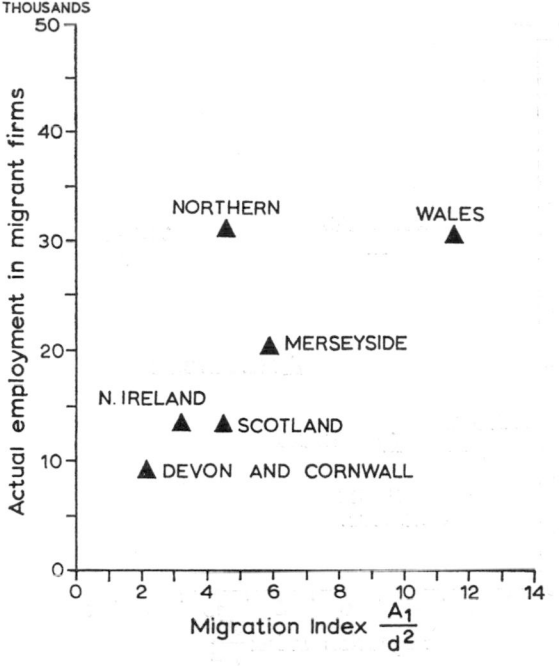

Figure 2.11. United Kingdom: predicted migration indices and actual migrant employment in the peripheral areas: movement from the Greater London Council area, 1945–65.
Source: see Table 2.2.

in no less than nine out of the eleven applications. In seven cases, the A_2 coefficients were higher. In other words, the output of the A_2 model does apparently fit the observed pattern of movement slightly better than the A_1 model, although in most cases an increase in the latter's distance exponent also improves the fit. The difference between the outputs of the A_1 and A_2 models is illustrated graphically for the GLC case ($b = 2.0$) in Figures 2.11 and 2.12.

These findings strongly support at an aggregate level the conclusions of micro-level studies noted earlier. In spite of the crudity of the data and models used, spatial variations in migration to the peripheral regions do seem significantly to be related to distance and labour availability, when these are combined within a simple gravity model. Moreover, modification of the 'attractiveness' variable by a regional market accessibility index, as suggested by the findings of micro-level analysis, generally improves the correlation still further. Of course, the simple approach used here is open to criticism on various grounds, not least that it fails to evaluate

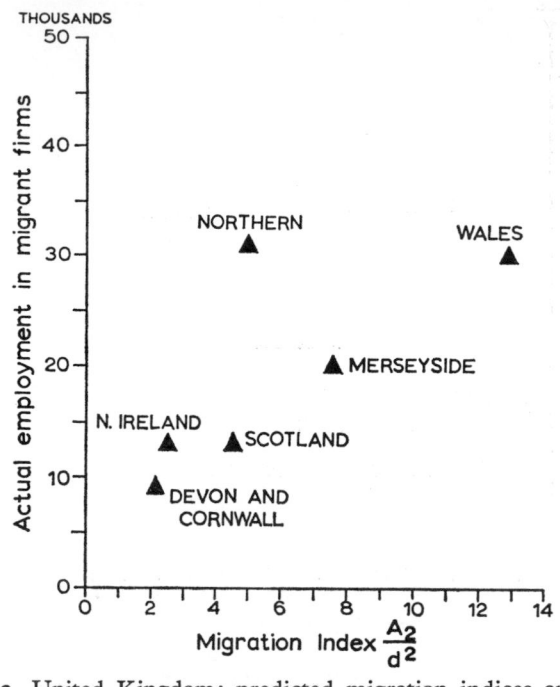

Figure 2.12. United Kingdom: predicted migration indices and actual migrant employment in the peripheral areas: movement from the Greater London Council area, 1945–65.

Source: see Table 2.2.

simultaneously movement from different origin areas, and hence allow for the effect of simultaneous real-world competition for labour resources between the different movement streams. It also ignores the not-inconsiderable temporal shifts which have occurred in the volume of movement to different destination regions. Nonetheless, the dominance of the basic distance and labour variables investigated here in explaining movement variations to different peripheral regions, independently attested by both micro- and macro-level analysis, must surely now be accepted.

The above analysis also suggests further intriguing questions for consideration. Why do the best-fit distance exponents vary between different areas of origin? More specifically, is there any significance in the fact that the friction of distance for movement from the South East and GLC area is apparently greater ($b = 2.5$) than that for movement from the West Midlands ($b = 1.5$)? And what additional factors, if any, might help explain such deviations from the average relationship as the unusually low value for movement from the West Midlands Conurbation to the Northern region (Figure 2.10), or the unusually high flow to the same area from the South East and London (Figures 2.9, 2.11 and 2.12)? Might the spatial arrangement of Britain's trunk road and rail networks have played a part here, in discouraging contact between the poorly-connected West Midlands and north-east England, while encouraging it along the well-established and high-speed A1 and east coast rail routes between London and Newcastle? Or do these deviations reflect differences in the orientation to different potential origin regions of the North East's local authority and Development Council's industrial promotion activity? As in many macro-level analyses, the present study thus also suggests interesting problems for further research.

The second example of macro-level analysis concerns intra-regional industrial overspill movement from London. Analysis at this equally-important but more detailed geographical scale is unfortunately greatly inhibited by the coarseness of the areal mesh in terms of which the Board of Trade's data are organised. The full movement matrix in fact breaks down the South East standard region outside London into only six areas (see Figures 2.7 and 2.8), the two largest of which (Essex, Hertfordshire and Bedfordshire; and Sussex and east and central Kent) account for no less than 63 per cent (118,000 jobs) of the total South East employment resulting from industrial migration from London. Moreover, the boundaries adopted by the Board of Trade are unfortunately by no means always the most meaningful or useful, in terms either of known patterns of industrial movement at a detailed scale or of boundaries used for the compilation of other official statistics. These problems of over-aggregation and boundary configuration unfortunately rule out attempts to test, for example, whether the unique national centrality (Caesar, 1964) and hence high economic potential (Clark, 1966) of the north-west quadrant of the region has attracted more movement to that sector than might be expected on other grounds. Even more important is the impossibility of throwing any light whatever on the vital question of distance–decay gradients for manufacturing movement from London, both as between different quadrants of the region and different industries or movement types.

However, the data do at least permit a crude macro-level test of two of the findings of micro-level analysis noted earlier, concerning radial migration and the significance of spatial variations in labour supply within the region. As pointed out above, micro-level evidence suggests that one of the stronger influences on migrant firm locational choice within south-east England has been the marked tendency towards radial migration outwards in each sector. Although this cannot be investigated for particular sectors, it was possible to group the Board of Trade's South East areas into two broad but fairly clearly defined northern and southern zones (see Figure 2.7), and to test the hypothesis that London industry already located north of the Thames has tended to migrate to the former, and that located to the south of the Thames to the latter. The key problem here was to arrive at some predictive index of movement from each London origin area, bearing in mind both the considerable difference in volume of industry in these two areas, and the different migration propensities of different London industries.

The method adopted utilised calculations already available as part of an SSRC (Social Science Research Council) sponsored research project on the location of recent industrial growth in South East England (Keeble and Hauser, 1970). Early post-war manufacturing employment in each of the two origin areas, broken down into major industry sub-totals, was obtained by aggregation of individual urban area estimates derived from the 1951 Census of Population. The needs of subsequent analysis enforced complicated adjustment of the industry sub-totals to allow for classificatory differences between the 1948 and 1958 Standard Industrial Classifications, and to derive estimates for the 14 industry groups recognised by the latter. Each of these industry estimates was then weighted by a mobility index, relating to the movement propensity of that industry. These mobility indexes were obtained by comparing the proportionate 1951 London manufacturing employment share of a given industry, with its share of subsequent migration (1952–65) to locations in the South East and East Anglia as defined by the employment estimates given in Appendix G of Howard's study (1968). The importance of this step is evident from the fact that mobility indexes varied from 0.000 (shipbuilding and marine engineering) and 0.214 (leather, leather goods and fur) to 1.613 (engineering and electrical goods) and 1.689 (chemicals and allied industries). Summation of each resultant mobility-weighted industry sub-total, for each of the two origin areas, thus yielded measures of the proportionate share of total actual migrant employment likely to be created in the two South East reception zones, on the assumption of radial migration north and south of the Thames. These measures, translated into predicted migrant employment, are given in Table 2.4, which also records the actual Board of Trade migrant employment estimates.

As the table shows, this simple radial migration (RM) model failed to yield migration estimates consonant with the observed values. The difference, as measured by the χ^2 test, is so great as to disprove the hypothesis that variations in industrial migration from London to reception areas

TABLE 2.4. *South East England: intra-regional movement, radial migration and labour availability model predictions, 1945–65*

Area	(A) Actual moves from G.L.C. area, 1945–65 (number)	(B) Actual employment in firms from G.L.C. area (thousands)	(C) RM model predicted employment (thousands)	(D) RML model predicted employment (thousands)
Northern (Essex, Herts, Beds, Bucks, Oxfordshire)	427	97	124	100
Southern (Kent, Sussex, Surrey, Hants and I.O.W.)	317	87	60	84
Total South East (exc. G.L.C. area)	744	184	184	184

Note: Columns B:C χ^2 = 18.029. Frequencies significantly different at 0.001 level.

Columns B:D χ^2 = 0.190. Frequencies not significantly different.

Sources: Howard (1968); unpublished DEP unemployment data; General Register Office, 1951 Census of Population.

north and south of the Thames can be explained solely by radial movement. Had this variable alone been operating, the northern zone should have received much more, and the southern zone much less, migrant industrial employment than they actually did. However, the RM model, though inaccurate in its predictions, at least anticipated correctly the direction of bias in observed movement towards the northern zone. It was therefore decided not to abandon a radial migration approach, but to modify it. Again, the direction of modification was suggested by micro-level studies. As already pointed out, these suggest that an important additional influence upon intra-regional movement streams in South-East England has been spatial variation in labour availability. The latter may crudely be measured by unemployment levels. Not only has this directly distorted radial movement flows in at least one case (east Kent), but it has greatly influenced Board of Trade policy on IDCs, which in turn has accentuated movement to South East coastal settlements suffering from relatively-high unemployment. Incorporation of this additional variable in the second (RML) model, however, involved the crucial step of quantification of its relative weight or importance, compared with radial movement.

Again, micro-level evidence suggested a logical basis for quantification. Processing of unpublished survey questionnaires relating to North-West London firms which established 56 migrant factories (transfers and branches) in the South East standard region outside London between 1940 and 1964 revealed that exactly one-third of all employment in these factories was provided by firms which acknowledged without prompting the importance of local labour availability in their locational decision. This group of labour-influenced intra-regional migrants deliberately excludes moves to new or expanded towns, and those to new locations close enough to North-West London to permit retention of an existing labourforce. Both these categories are however included in the total employment calculation. Clearly, extrapolation of this sample estimate as a measure of the proportion of total London migrant employment likely to have been influenced by geographical variation in labour availability within the region is based on assumptions which are easy to question. But in the absence of any other logical basis for estimation, this approach was accepted as reasonable. In the RML model, therefore, two-thirds of total predicted movement from London was allocated zonally on the original basis of radial migration, while one-third was allocated on the basis of labour availability. The latter was measured by the average annual unemployed workforce in each zone over the three-year period 1958–60. The difference in this respect between the northern and southern zones was in fact considerable, the former possessing only 8,760 unemployed workers compared with the latter's 21,770. Summation of the separate employment estimates thus derived from the radial migration and labour availability components of the RML model yielded the overall model prediction for the two reception zones recorded in Table 2.4.

This prediction, unlike that of the single-variable RM model, is remarkably accurate. Indeed, the χ^2 test confirms that prediction and reality are so close as to be statistically indistinguishable. Although the northern zone prediction is still slightly higher, and the southern zone value slightly lower, than the observed totals, the discrepancy is remarkably small bearing in mind the sample extrapolation procedure used, problems of industrial classification adjustment, and other sources of data inaccuracy. It would of course be tempting to try and refine the result still further, by suggesting additional influences which might explain the residual error term. Such factors as the concentration of high-amenity residential areas in the southern zone, or the exclusion from this calculation of the small amount of movement (13,000 jobs) to East Anglia and Northamptonshire, spring to mind. But such refinement is not justified, given the closeness of fit and the inevitability of small random errors in the model's output as a result of data inadequacies. In sum, simple macro-level analysis strongly supports micro-level findings concerning the importance of radial migration, and to a lesser extent of local labour (and IDC) availability, in influencing the geography of intra-regional industrial movement from London to surrounding parts of south-east England. A corollary of this finding is of course the implication that at least on the very aggregate geographical scale for which movement data are available, other postulated locational influences, such as the distribution of new towns or of attractive residential areas, are probably of only very minor, if any, importance.

Conclusions

The preceding discussion confirms that employment mobility is of major importance both for an understanding of the changing spatial pattern of Britain's economy, and for national, regional and local planning. Not only has the scale of post-war movement, at least in the manufacturing sector, been considerable, but mobile manufacturing firms have tended to be unusually virile, large and export-orientated. The bulk of such movement has conformed to one or other of two distinct migration types, defined by distance of move, nature of destination area, and migrant establishment size and organisational tendencies. Preceding discussion has also drawn attention to the possible use of local office linkage measures, and manufacturing marketing orientation, as indices of potential mobility; to the advantages and potential of macro-level analysis in mobility studies; and to the remaining need, despite Howard's invaluable study (1968), for comprehensive and above all disaggregated movement data, both of office migration generally, and of manufacturing movement at the detailed intra-regional and inter-industry scales.

On the other hand, space limitations have prevented discussion of many other important questions raised by post-war employment mobility. What, for example, are its implications for the traditional twentieth century centre–periphery dichotomy in the geography of economic health and growth in Britain (Caesar, 1964)? After all, Howard's data (1968) indicate that Britain's non-peripheral areas, nearly all of which can be classified in Caesar's central (or 'Midlands') growth zone, have received more UK-generated mobile manufacturing employment (399,000 jobs) and far more of the self-sufficient transfer moves (901 establishments) than the peripheral areas (363,000 jobs and only 200 establishments, respectively). Has the post-war infusion of mobile growth industry to the lagging periphery really begun to engender there a self-sustaining, Myrdalian-type, cumulative economic growth process? Or are the continuing and in many ways strengthening advantages of the centre (Caesar, 1964; Keeble, 1972) too powerful to allow for peripheral growth which is more than a reluctant tribute to government controls, or less than fundamentally dependent upon the dynamism of the economic heartland?

What too of the cost/benefit ratio of government employment mobility policy, in view of the recent remarkable increase in financial inducements paid to manufacturing firms expanding at the periphery (from £17 million in 1962–3, to approximately £260 million in 1968–9)? Is there a real conflict between the industrial needs of the peripheral regions, and those of new central overspill communities, such as Milton Keynes and Telford, given limitations on the volume of mobile industry at any one time? Ought peripheral area industrial location strategy to take more account of the currently fashionable concepts of growth centres, and industrial complex analysis, as recently applied for the first time in Britain to the problem of industrial mobility programming for the Leyland–Chorley new city of central Lancashire (Economic Consultants Ltd, 1969)? Should London and Birmingham continue to be regarded as unfailing and unaffected sources of migrant enterprise, and properly subject to stringent industrial and office controls designed to prise growing firms from a metropolitan seed-bed? Does not recent office decentralisation experience, discussed above, demonstrate the exceptional strength of the functional bonds tying most central-London office firms to the capital, and the fallacy of an inter-regional office mobility policy based on an analogy with manufacturing industry? And do not national economic considerations and the onset of accelerating manufacturing employment decline in Greater London, for the first time in its history, support the Greater London Council's plea for a more selective IDC policy based on careful identification of London industries with above-average productivity levels? These major policy questions cannot possibly be answered adequately here. But their range and significance bear witness to the importance of employment mobility in the context of local, regional and national planning over the next few decades, and to the pressing need for research into the incidence, character and spatial implications of the movement of manufacturing and service activity in Britain.

References

Beacham, A. and W. T. Osborn (1970). 'The Movement of Manufacturing Industry', *Regional Studies*, IV, 1, 41–7.

Brown, C. M. (1965). *Industry in the New Towns of the London Region: a study in industrial decentralisation*, unpublished MSc(Econ.) thesis, University of London.

Caesar, A. A. L. (1964). 'Planning and the Geography of Great Britain', *Advancement of Science*, XXI, 230–40.

Cameron, G. C. and B. D. Clark (1966). *Industrial Movement and the Regional Problem*, Oliver and Boyd, University of Glasgow Social and Economic Studies, Occasional Papers No. 5.

Cameron, G. C. and K. M. Johnson (1969). 'Comprehensive Urban Renewal and Industrial Relocation – The Glasgow Case', Chapter 10 in J. B. Cullingworth and S. C. Orr (eds.), *Regional and Urban Studies*, Allen and Unwin.

Clark, C. (1966). Industrial Location and Economic Potential, *Lloyds Bank Review*, 82, 1–17.

Daniels, P. W. (1969). 'Office Decentralisation from London – Policy and Practice', *Regional Studies*, III, 2, 171–8.

Daniels, P. W. (1970). 'Employment Decentralisation and the Journey to Work', *Area*, 3, 47–51.

Dowie, R. (1968). *Government Assistance to Industry: a Review of the Legislation of the 1960s*, Centre for Research in the Social Sciences, University of Kent, Ashford Study Paper II.

Dunning, J. H. (1969). 'The City of London: A Case Study in Urban Economics', *Town Planning Review*, XL, 3, 207–32.

Economic Consultants Ltd (1969). *Study for an Industrial Complex in Central Lancashire*, unpublished study for Department of Economic Affairs.

Economist Intelligence Unit Ltd (1964). *A Survey of Factors Governing the Location of Offices in the London Area*, Location of Offices Bureau.

Goddard, J. (1970). 'Greater London Development Plan. Central London: A Key to Strategic Planning', *Area*, 3, 52–4.

Haig, R. M. (1926). 'Towards an Understanding of the Metropolis', *Quarterly Journal of Economics*, XI, 2, 179–208; 3, 402–34.

Hall, R. (1970). 'An estimate of office decentralisation', unpublished paper read to Location of Offices Bureau seminar, 30 June.

Hammond, E. (1964). 'The Main Provincial Towns as Commercial Centres', *Urban Studies*, I, 2, 129–37.

Hammond, E. (1967). 'Dispersal of Government Offices: A Survey', *Urban Studies*, IV, 3, 258–75.

Howard, R. S. (1968). *The Movement of Manufacturing Industry in the United Kingdom 1945–65*, HMSO for the Board of Trade.

Hunker, H. L. and A. J. Wright (1963). *Factors of Industrial Location in Ohio*, Columbus, Ohio, Bureau of Business Research, Ohio State University.

Keeble, D. E. (1965). 'Industrial Migration from North-West London, 1940–1964', *Urban Studies*, II, 1, 15–32.

Keeble, D. E. (1968). 'Industrial Decentralization and the Metropolis: the North-West London Case', *Transactions*, Institute of British Geographers, XLIV, 1–54.

Keeble, D. E. (1969). 'Local Industrial Linkage and Manufacturing Growth in Outer London', *Town Planning Review*, XL, 2, 163–88.

Keeble, D. E. (1972). 'The South East and East Anglia', Chapters 2 and 3 in G. Manners (ed.), *Regional Development in Britain*, Wiley.

Keeble, D. E. and D. P. Hauser (1970). *Spatial Correlates of Industrial Growth in Outer South-East England*, Social Science Research Council Final Report.

Law, D. (1964). 'Industrial Movement and Locational Advantage', *Manchester School of Economic and Social Studies*, XXXII, 2, 131–54.

Lewis, E. W. (1971). *The Location of Manufacturing Industry in the Western Home Counties*, unpublished M. Phil. thesis, University of London.

Loasby, B. J (1967). 'Making Location Policy Work', *Lloyds Bank Review*, 83, 34–47.

Location of Offices Bureau (1970). *Annual Report, 1969–70*, LOB.

Luttrell, W. F. (1962). *Factory Location and Industrial Movement: A Study of Recent Experience in Great Britain*, National Institute of Economic and Social Research, 2 vols.

McCrone, G. (1969). *Regional Policy in Britain*, Allen and Unwin.

Royal Commission on the Distribution of the Industrial Population (1940). *Report* (Chairman Sir Montague Barlow, Cmd. 6153), HMSO.

Seaman, J. M. (1970). *The Location of Manufacturing in South East London: an Industrial Geography*, unpublished MSc(Econ.) thesis, University of London.

Secretary of State for Economic Affairs (1969). *The Intermediate Areas* (Report of a Committee under the Chairmanship of Sir Joseph Hunt, Cmnd. 3998), HMSO.

Smith, B. M. D. (1970). 'Industrial Overspill in Theory and Practice: the Case of the West Midlands', *Urban Studies*, VII, 2, 189–204.

Smith, T. R. (1954). 'Locational Analysis of New Manufacturing Plants in the United States', *Tijdschrift voor Economische en Sociale Geografie*, XLV, 2, 46–50.

South East Joint Planning Team (1970). *Strategic Plan for the South East*, HMSO.

Spooner, D. J. (1970). *Industrial Development in Devon and Cornwall 1939–67*, unpublished typescript.

Stevens, B. H. and C. A. Brackett (1967). *Industrial Location: A Review and Annotated Bibliography of Theoretical, Empirical and Case Studies*, Regional Science Research Institute, Philadelphia, Bibliography Series No. 3.

Townroe, P. M. (1971). *Industrial Location Decisions: A Study in Managerial Behaviour*, University of Birmingham, Centre for Urban and Regional Studies, Occasional Papers No. 15.

Twyman, P. H. (1967). *Industrial Location in East Kent, with Special Reference to Factory Relocation and Branch Plant Establishment*, unpublished MSc(Econ.) thesis, University of London.

Wabe, J. S. (1966). 'Office Decentralisation: An Empirical Study', *Urban Studies*, III, 1, 35–55.

3. Leads and lags in inter-regional systems: a study of cyclic fluctuations in the South West economy

PETER HAGGETT

Although the death of the business cycle has been frequently proclaimed in the post-Keynesian era there are grounds for thinking that, like Mark Twain's obituary notice, such claims are a little premature. For although the post World War II years have seen no economic or financial crashes on the scale of 1929, a large number of statistical indicators continue to show wave-like patterns of between two and twelve years' duration – the conventional range of business cycles – superimposed on longer-term aggregate growth (Bronfenbrenner, 1969). For the United Kingdom economy as a whole, four major cycles in industrial production have been recognised by the Central Statistical Office in the last two decades. The longest extended over a seventy-three month period from 1952–VII to 1958–VIII with a minor recession in the summer of 1956. This was succeeded by two fifty-three month cycles (1958–VIII to 1963–I and 1963–I to 1967–VI). The latest period falls within the fourth cycle which to January 1970 had lasted for thirty months. As Figure 3.1 shows, the cycles of industrial production are matched by cycles of unemployment although here the cycle lags some months behind the production cycle. This lag has varied from four months at the start of the first cycle to three months for the start of the second cycle and two months for the third cycle. The last cycle showed a curious double peak in the summers of 1967 and 1968. The start given to the present unemployment cycle is therefore perhaps rather arbitrary and as currently defined lags thirteen months behind the start of the production cycle.

This chapter restricts its examination to fluctuations in one economic indicator (the number registered as unemployed expressed as a percentage of the workforce) for one major planning region (the South West Economic Planning Region) over varying periods within the last two decades. Statistical relationships between regions are examined at four distinct geographical levels: (1) the *inter-regional* level with comparisons between the South West region and other major regions; (2) the *inter-metropolitan*

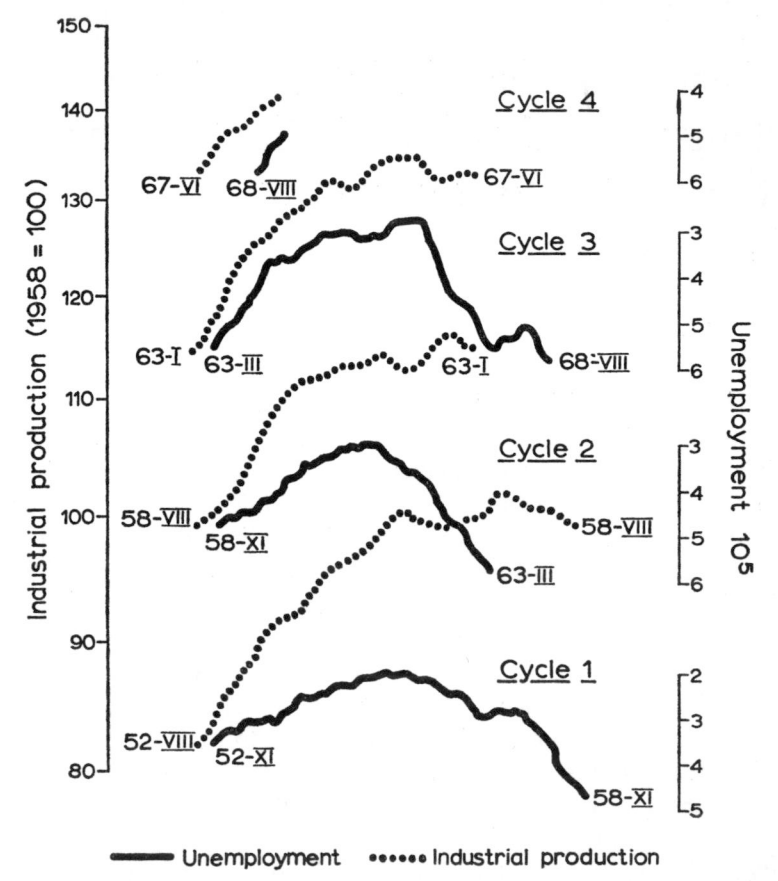

Figure 3.1. Great Britain: cycles of industrial production and cycles of unemployment, 1952–69.

Note that the figures for wholly unemployed are seasonally adjusted and that the scale for unemployment has been inverted to allow comparison of the two trends. The unemployment cycle lags some months behind the production cycle.

Source: Central Statistical Office, 1969.

level with comparisons between the two leading south-western cities, Bristol and Plymouth; (3) the *inter-area* level with comparisons between sixty areas within the South West region; and (4) the *urban-hierarchy* level with comparison between Bristol and smaller urban centres in the Bristol sub-region. The results of the four levels are compared and ways of using the findings are examined.

Unemployment rates as indicators of cyclic fluctuations

The single indicator used in this series of comparison was the percentage unemployment rate. One major advantage of unemployment statistics from both the spatial and the dynamic viewpoint is that figures are published at monthly intervals for a rather close network of local areas (Figure 3.2). Such networks are subject to some change over time, e.g. for the period under study within the South West (July 1960 to December 1969) the number of local employment exchanges has varied and some grouping is necessary to maintain comparability over long time periods. To go back before July 1960 would have meant a progressive reduction through merging in the number of comparable areas. The fineness of the spatial and temporal graticule and its sensitivity to cyclic changes made it superior to other indicators considered, e.g. housing starts and activity rates. As Prest (1968, 21) argues, unemployment data now serve as 'the best single indicator of cyclical performance in the post-war period'.

The unemployment rate has, however, been attacked by Eversley (1968) as a poor indicator on the grounds that (1) it misallocates the unemployed between regions since statistics relate to the registered unemployed in the area where their insurance cards were exchanged and are only partly adjusted for the fact that this may not be the same area as the one in which they work; (2) it takes no account of the number deterred from registering at the local unemployment exchange by the local job shortage (this principally refers to women); and (3) it ignores the outward movement of job seekers to look for work in other areas. However, Weltman and Rendel (1968) criticise Eversley's alternative 'true measure of unemployment' (which includes estimates of migrant males) on the ground that it represents only one of a large number of possible indices. The lack of correlation between such indices and the absence of agreement on any alternative measure of unemployment suggest that the existing 'crude' rate might safely be used.

Leads and lags at the inter-regional level

A major analysis of quarterly unemployment percentages for ten British regions was published by Brechling in 1967. Since this includes the South West it enables direct comparison of fluctuations within the region as a whole and other major regions (Figure 3.2).

Brechling's (1967) model divides the unemployment level of any sub-region into three components:

$$U_{jt} = A_{jt} . S_{jt} . R_{jt}$$

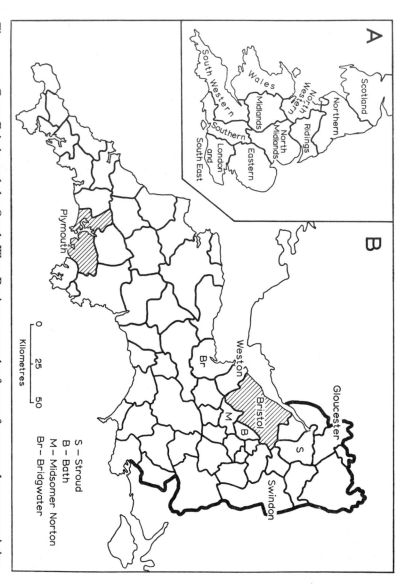

Figure 3.2. Great Britain and the South West Region: network of areas for unemployment statistics. (A) Standard regions. (B) Local employment office districts.
Note: some of the districts in the South West Region have been amalgamated to preserve comparability of records over the period 1960–69.

where U_{jt} is the local level in the jth sub-region; A_{jt} is the *aggregative cyclic component* which reflects the level of economic activity of the country as a whole; S_{jt} is the *structural component* peculiar to the jth region and is regarded as either constant or changing over time; while R_{jt} is the *regional cyclic component* caused by the particular industry-mix of the sub-region or its local industrial character (e.g. proportion of externally-owned plants).

The three components are combined by Brechling into a single equation:

$$U_{jt} = \alpha_j\, U_N{}^{\beta_{1j}}(t+n_j)e^{\beta_{2j}t+\beta_{3j}t^2}\, R_{jt},$$

where U_N is the national unemployment level, t is time, n_j is the length of lag or lead (and may be either negative or positive), α_j is the level of structural parameters in period t_0 and the betas are coefficients. β_{1j} describes the elasticity of the aggregative cyclic component in the jth sub-region with respect to the national level (U_N), while β_{2j} and β_{3j} describe changes in the structural component with respect to time. The time trend may be quadratic in form to allow for the possibility that the structural component changes at an accelerating or decelerating rate.

For computing purposes, the equation may be handled as either a multiplicative (log-linear) or additive (arithmetic-linear) structure. The appropriate equations are:

$$\log \hat{U}_{jt} = \log \alpha_j + \beta_{1j}\log U_{N(t+n_j)} + \beta_{2j}\log t + \beta_{3j}\log t^2$$

for the log-linear model and

$$\hat{U}_{jt} = a_j + b_{1j}U_{N(t+n_j)} + b_{2j}t + b_{3j}t^2$$

for the arithmetic-linear model. In each case, the regional cyclic component is given as a residual:

$$\log R_{jt} = \log U_{jt} - \log \hat{U}_{jt}$$

or

$$R_{jt} = U_{jt} - \hat{U}_{jt}$$

in the log-linear and arithmetic-linear cases respectively. Proportionate changes in employment may give a more relevant index than absolute changes in some areas than others, but there seems to be no hard and fast rule to decide between the two alternative models. Brechling opts for the log-linear structure on the ground that the elasticity yielded (i.e. β_1 coefficients) is more easily interpreted in economic terms.

The findings of Brechling's work on inter-regional leads and lags are shown in Figure 3.3, based on quarterly unemployment percentages for the period 1952-III to 1963-II (forty-four time periods) for ten British regions. Regression equations were run for each region against national levels for assumed lag values of $n_j = -2, -1, 0, +1,$ and $+2$ for both the log-linear and arithmetic-linear models. For seven of the ten regions, the best fit was obtained with $n_j = 0$. A lead of one-quarter gave the best fit for the Midlands and a lag of one-quarter the best fit for Scotland and the North. As Figure 3.3 shows, broadly comparable results were obtained with both models. Figure 3.3A shows profiles of coefficients of determination (R^2) indicating that even where the closest fit was given with $n_j = 0$ there was sufficient asymmetry in correlations for preceding and following quarters to suggest that data for periods finer than one quarter might have revealed slight lags. The asymmetry in the profiles is indicated in Figure 3.3B, which plots the median point for the five R^2 values. Wales lies very near the centre for both models but the other regions show some evidence of lags or leads that are consistent in direction with the results obtained by varying the discrete lags. On the evidence of Brechling's results, the South West would appear to lie rather close to the national average but with a slight tendency to lead rather than lag.

Leads and lags at the inter-metropolitan level

Comparison of fluctuations between major cities has been less commonly conducted than comparisons between major regions. Within the South West, Bassett and Tinline (1970) have compared fluctuations in quarterly unemployment levels between the Bristol and Plymouth sub-regions over the post-war period (1947–69). They adopt the method of spectral analysis (Granger, 1964) by which the unemployment series can be decomposed into a number of cyclical components of different frequencies under assumptions of stationarity. Estimates of spectral density for the two series (Figure 3.4B) show peaks at 5.5 years (roughly approximating the mean wavelength of the post-war business cycle), at 1.0 year (the seasonal wavelength), and evidence of a weaker peak at 2 years. The seasonal effect is stronger at Plymouth than Bristol, reflecting the relative importance of tourism and related industries in the two areas.

Cross-spectral analysis is used to analyse fluctuations in the two series simultaneously. The extent to which any frequency component in one series can be correlated with the same frequency component in the other series can be estimated. The plot of the correlation against wavelength in a 'coherence diagram' (Figure 3.4C) shows an average coherence between the two series of 0.59, with peaks corresponding to the matching waves at 5.5, 2 and 1 years as identified above. Although successive frequency components may be correlated they may lag behind each other, so that cycles in one series on a given wavelength are followed by cycles of a similar

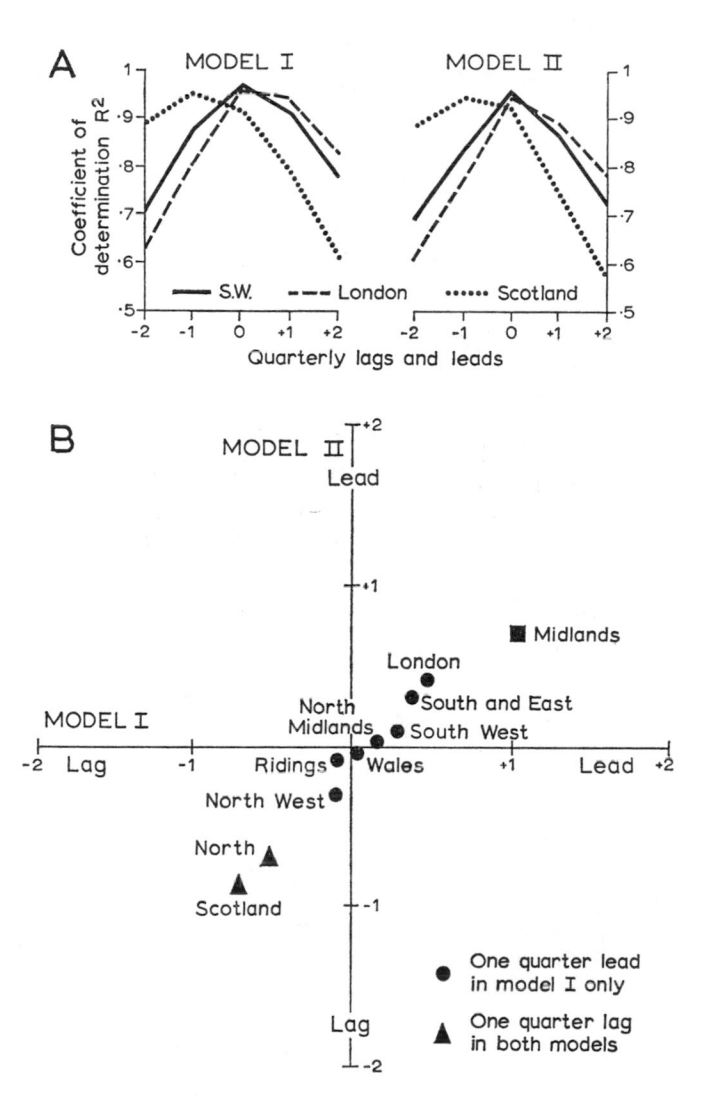

Figure 3.3. United Kingdom: quarterly unemployment data, 1952–63. (A) Correlation coefficients for the relationship of total UK unemployment levels to regional unemployment levels for various leads and lags. (B) Estimated leads and lags for log-linear model (I) and arithmetically-linear model (II).

Source: Brechling, 1967, 8.

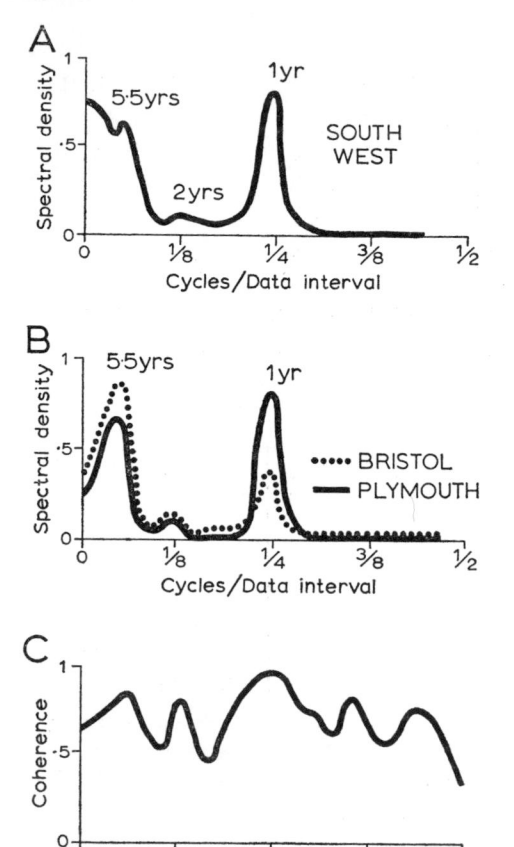

Figure 3.4. South West Region: spectral analysis of quarterly unemployment levels, 1947–69. (A) Whole region. (B) Bristol and Plymouth sub-regions. (C) Coherence of the Bristol and Plymouth series.
    Source: Bassett and Tinline, 1970, 20.

wavelength. This phenomenon may be measured in spectral analysis as 'phase lags' and plotted against wavelength in the same manner as the coherence diagrams. No significant lags between Plymouth and Bristol over the post-war period were revealed by the phase angle diagrams.

Spectral analysis clearly provides a powerful method for the comparison of regional time series. Users of the method have to meet a number of rigorous assumptions (Granger, 1964) and the validity of estimates for both the existence of frequency components and their phase lags is clearly related to the length of the time series record available. The failure to find significant lead–lag relationships between the two leading metropolitan areas in the South West may partly relate to the use of quarterly data. The possibility of detecting shorter lags using monthly data is taken up in the next two sections.

Leads and lags at the inter-area level

A study which utilised the fine-grained quality of unemployment data was published by Bassett and Haggett in 1971. Series for sixty local areas were analysed over 114 consecutive months from July 1960 up to and including December 1969. The period chosen was selected so as to be (1) long enough to meet at least the minimal requirements of spectral analysis (i.e. 100 or more equally-spaced observations), (2) long enough to include some major cycles but (3) short enough to avoid major changes in local employment-district boundaries.

The variability of the sixty series shows considerable contrasts. The most extreme fluctuations are associated with coastal resort centres such as Newquay. The least variation was shown by small- to medium-sized areas with a broad spectrum of service industries and stable manufacturing industry (e.g. Street). Figure 3.5 shows the overall regional trends in the spatial pattern of unemployment for a recent month (December 1969) and regional trends in the distribution of total variation within the South West region. One characteristic feature of unemployment series is their seasonal variation and Figure 3.6 shows arrays of values for five representative areas. About half of the sub-areas have more than fifty per cent of the total variance made up of seasonal and irregular elements. This figure is as high as 93.2 per cent and 91.0 per cent in the case of coastal resort areas (Penzance in West Cornwall and Ilfracombe in North Devon). There is also some evidence that seasonal variations may have damped in the latter half of the study period, possibly in response to the Selective Employment Tax introduced in September 1966 (see for example the findings of Reddaway, 1970).

The time series for each of the sixty sub-areas in the South West region was filtered in an attempt to remove these seasonal characteristics. A centred twelve-month moving average filter was used, of the form:

$$\hat{U}_t = \frac{U_{t-6}}{24} + \frac{U_{t-5}}{12} + \frac{U_{t-4}}{12} + \ldots \frac{U_{t+4}}{12} + \frac{U_{t+5}}{12} + \frac{U_{t+6}}{24}$$

where $\hat{U}_t$ is the filtered unemployment rate for month t, and U_{t-6} is the unemployment rate for the $(t-6)$th month, and so on. Bassett and Haggett (1971, 395) also point out while the filter effectively removes seasonal and

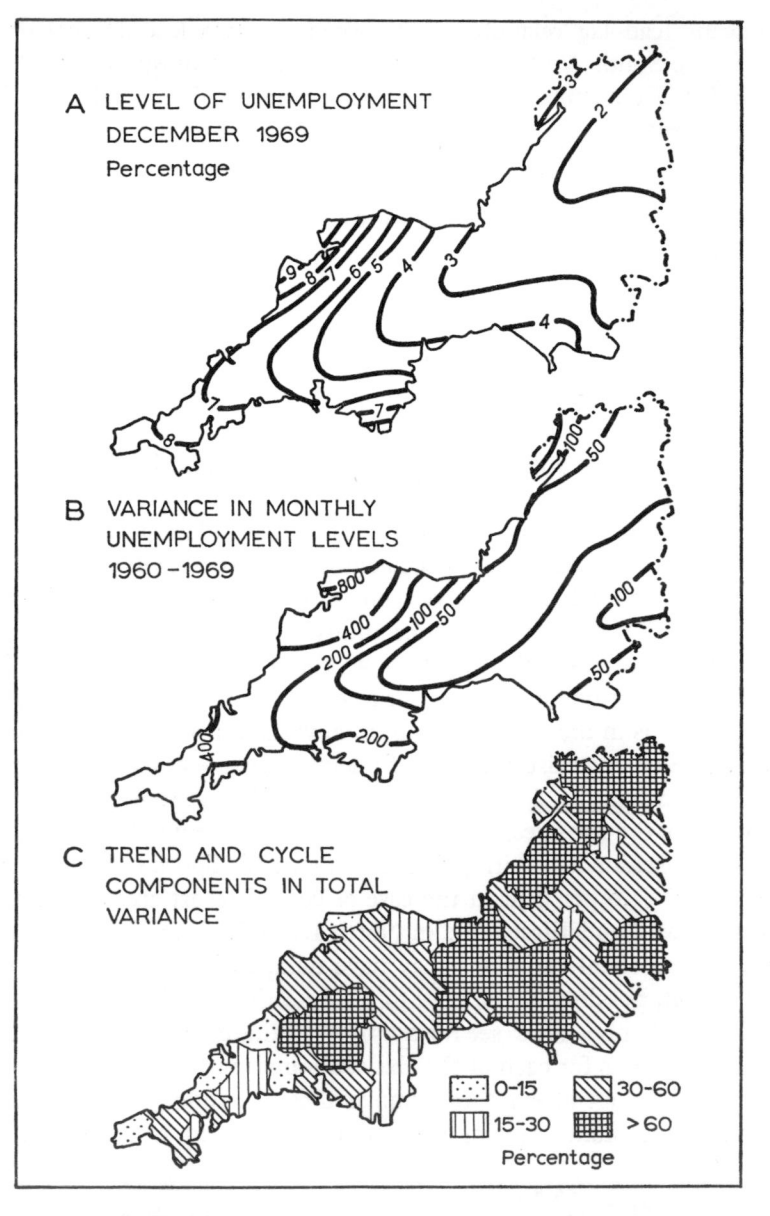

Figure 3.5. South West Region: unemployment, 1960–9. (A) Spatial variation in unemployment percentage levels at December, 1969. (B) Spatial variation in the total variance of unemployment levels, July 1960 to December 1969. (C) Percentage contribution to the total variance of the trend and cycle components, July 1960 to December 1969.

Note: maps (A) and (B) are based on trend-surface analysis of district values.

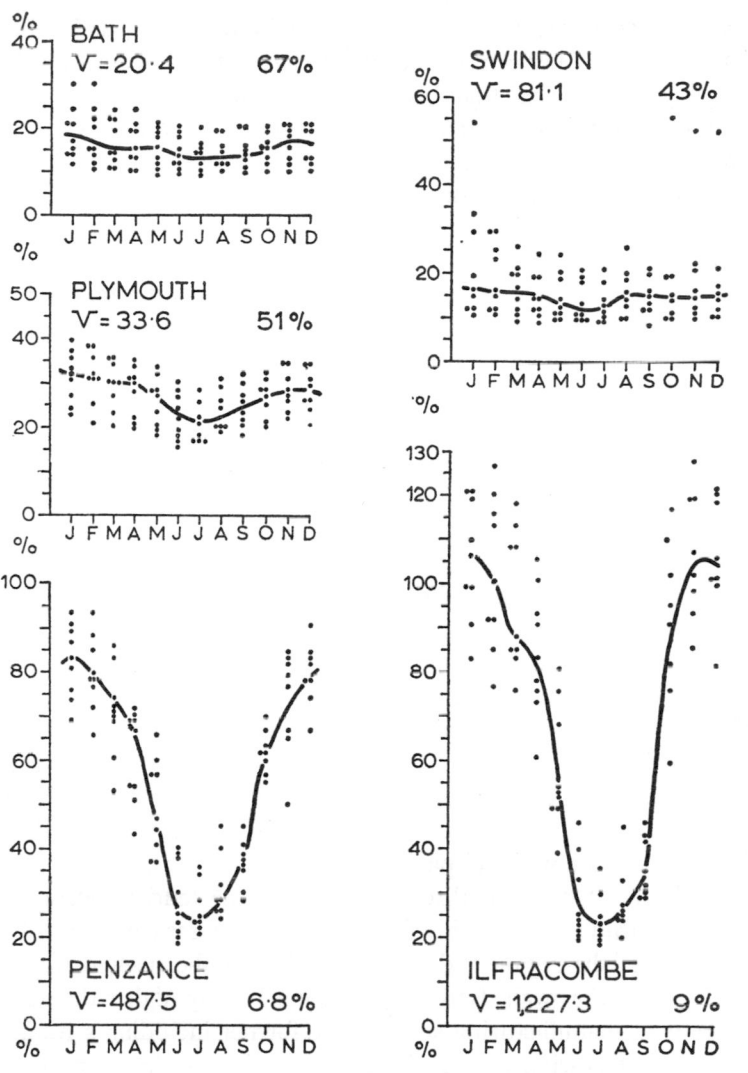

Figure 3.6. South West Region: arrays of monthly unemployment values for representative districts, 1960–9. The total variance (V) and the percentage thereof accounted for by trend and cycle elements (%) are shown.

higher frequencies from the series it also dampens to some extent some frequencies lower than the seasonal. It leaves in, however, the residual longer cycles which are the object of comparison. Within these residual cycles, four significant points were identified, i.e. (1) *Recovery mid-point* defined as the median point in any run of consecutive months in which the difference between the unemployment rate in preceding and following months drops by 0.1 per cent or more; (2) *Peak* defined as any month with an unemployment rate less than both the preceding and following months; (3) *Downturn mid-point* defined as the median point in any run of consecutive months in which the difference between the unemployment rate in preceding and following months rises by 0.1 per cent or more; (4) *Trough* defined as any month with an unemployment rate greater than both the preceding and following months. In practice, some recovery and downturn phases were as short as one month and in these cases the mid-point was defined as that month; for peaks and troughs lasting two or more months, the median point of the peak or trough was recorded.

Comparison was carried out of the timing of the four critical turning points for the sixty sub-areas. Bristol was chosen as a reference datum as it dominates the employment situation in the South West. The turning points identified for Bristol show a peak in June 1961 with the downturn mid-point in July 1962 leading to a major trough in April 1963. The mid-point for the recovery phase in September 1969 leads to a weakly defined peak in December 1964. A downturn in the following year (October 1966 leads to a trough in July 1967. No clear turning points on the series were recognised in 1968 or 1969. The spatial distribution of leads and lags for sub-areas in relation to Bristol at seven distinctive phases in the cycle fail to reveal strong sub-regional regularities (Figure 3.7). For most phases and for the majority of sub-areas the local turning points are within the same quarter as the regional capital; in the case of the sharp upturn of Autumn 1963 the dominance of 'Bristol behaviour' is almost complete. This reflects Bristol's close coincidence with the national pattern at this time. Sub-areas which show consistent leads of between three and six months tend to occur in two clusters: (1) within a ring roughly forty-eight kilometres from Bristol in the northern part of the region and (2) within the Camborne-Redruth-Truro triangle in West Cornwall. Both areas are marked by higher than average employment in manufacturing sectors, the former including a number of specialist manufacturing centres (e.g. Street with its dominant footwear industry) and the latter with considerable amounts of new light industry introduced under Government Development Area policies. Sub-areas showing distinctive lags are less common. The peripheral rural areas of north and central Devon and of south Somerset and Dorset show indications of falling into this class.

The method used by Bassett and Haggett was essentially a preliminary one. It used only extreme values in the series and utilised only seven of the 114 values in the series. Ambiguities in recognising turning points prevent

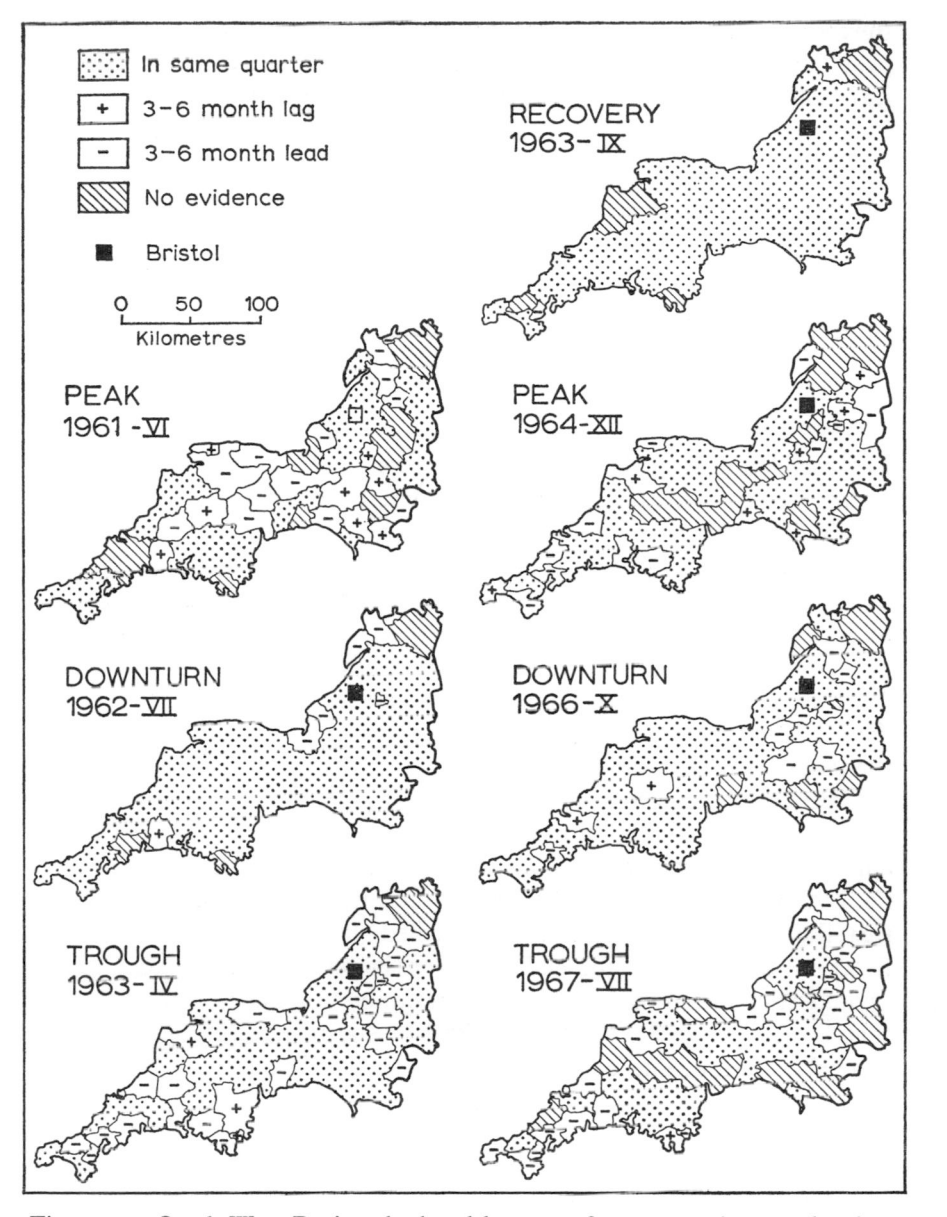

Figure 3.7. South West Region: lead and lag areas for seven points on the sixty unemployment trend-cycle series, Bristol as datum.
Source: Bassett and Haggett, 1971, 398.

the recognition of consistent 'leaders' or 'laggers' at the sub-regional level over the period studied. However the tentative patterns revealed suggest just enough regularity to warrant following up at a more detailed level.

Leads and lags at the local urban-hierarchy level

Bassett and Haggett (1971) followed up their preliminary study of the sixty sub-regions within the South West with a more detailed examination of localities within the Bristol area. In addition to Bristol itself, seven local sub-regions are included. The two methods used, lag correlation and spectral analysis, are reported separately.

Lag correlations

Lag correlation analysis computes the cross correlations between pairs of unemployment series at various time lags. Even after the seasonal and higher frequencies have been removed by filtering, high correlations between time series can be obtained simply because the series are trending in a similar way. Linear regression has therefore been used approximately to remove any trends so that the series that are correlated represent predominantly cyclical phenomena. Higher order polynomials could have been used to make a more accurate estimate of the trend, but there is a danger here that a polynomial curve may remove as a trend effect some of the long-wavelength cycles in which we are interested.

Cross correlations were calculated for lag values from -6 to $+6$ months. The curve for the correlation between Bristol and Weston-Super-Mare is approximately symmetrical, reaching a maximum with a lag value of zero, i.e. both series have the highest correlation with no lag displacement, so that the series are approximately matched. By contrast, the correlation function between Bristol and Swindon is generally lower and highly asymmetric. The profile reaches its maximum at a lag of -6, indicating that Swindon appears to 'lead' Bristol by six months (Figure 3.8). The general pattern of intercorrelations is shown by Figure 3.12.

Cross-spectra

The methods of cross-spectral analysis used in the intermetropolitan level (Bassett and Tinline, 1970) were extended by Bassett and Haggett (1971, 405–9) to the eight areas in the Bristol region. The original series were corrected for stationarity by removing linear trend but were not seasonally adjusted.

Figure 3.9 shows spectral estimates for relations between (1) Bristol and

BRISTOL LAGS

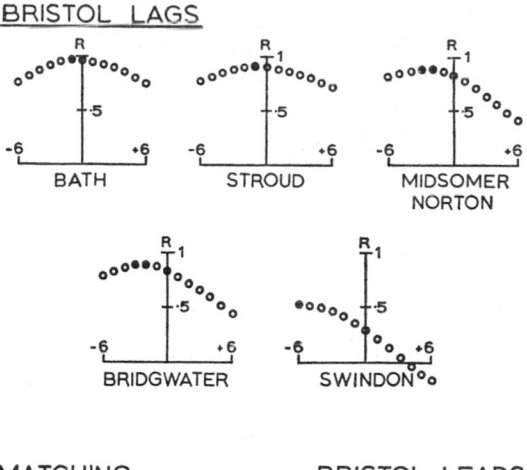

MATCHING   BRISTOL LEADS

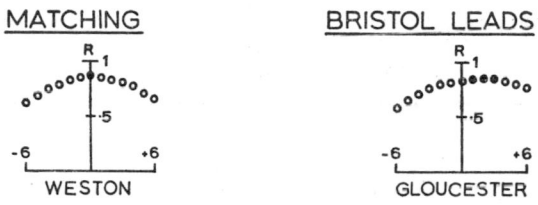

Figure 3.8. South West Region: cross-correlation values for pairs of unemployment series in the Bristol region, at various leads and lags relative to Bristol.
   Note: maximum values of the correlation coefficients are shown by solid points:
leads (+) and lags (−) are shown in months.
   Source: Bassett and Haggett, 1971, 402.

Bath and (2) Bristol and Swindon, using the same conventions as in the
Bristol–Plymouth intermetropolitan study. In the case of Bristol and Bath,
both spectra simply show a continuous and sharp decline from zero frequency.
The coherence spectrum shows a high relationship between frequency
components in the Bristol series and components of the same frequency in
the Bath series. The phase spectrum shows no evidence for the frequency
components of the Bath series to lead or lag the Bristol series.

   By contrast, the Swindon spectra show a marked difference from the
Bristol spectra in both long- and short-wave components. This difference
is reflected in the cross-spectral comparisons. The relationship between the
frequency components is shown as much lower in the coherence diagram,
while the phase angles show that the Swindon series leads Bristol very
strongly with respect to all frequency components up to the seasonal
frequency. Averaging the phase angles over the first four frequency points
(this is essentially the frequency band over which the cross correlations of

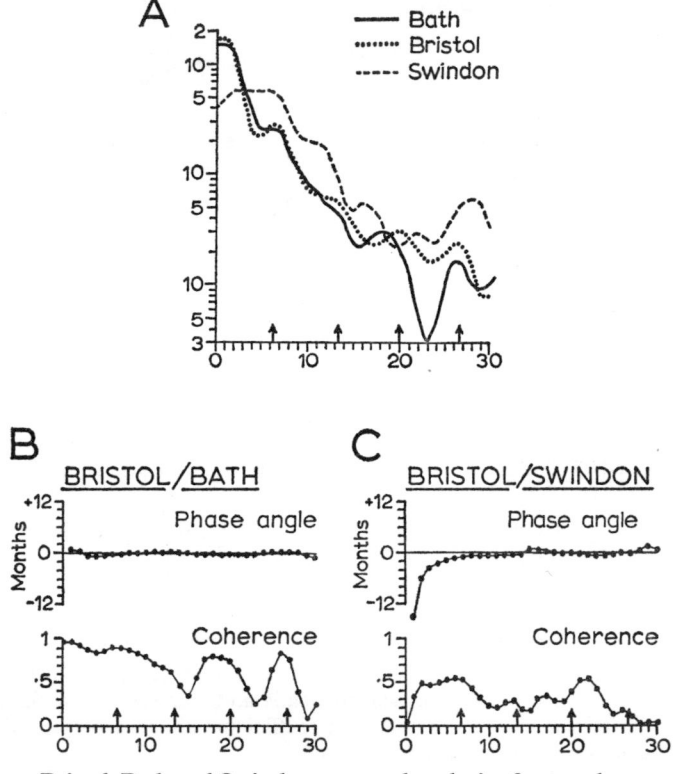

Figure 3.9. Bristol, Bath and Swindon: spectral analysis of unemployment series (A) and coherence profiles of the two pairs of series (B and C).
Source: Bassett and Haggett, 1971, 406–9.

the previous section were calculated) values are obtained which seem to confirm the relationships identified in Figure 3.8. That is, Bristol, Bath and Weston-Super-Mare emerge as a group roughly in phase, Gloucester lags slightly behind Bristol, and Swindon appears to vary somewhat independently.

The limitation of the lag-correlation and cross-spectral models with data for only 116 data points is discussed at length in the original paper (Bassett and Haggett, 1971, 409ff.). The consistency between the results obtained by the two methods using different types of trend removal suggest that some confidence may be placed in the results at least for the time-period studied. More fundamental objections may be raised to the use of the classical product–moment type of correlation coefficients when used for relating time series for one region in terms of another. For example, Theil (1958, 31) has pointed out its disadvantages in that perfect positive correlation ($r_{xy} = +1.00$) does not imply perfect forecasting, but only an exact linear relationship with a positive slope between the changes of activity in one region (X_i) and changes in another (Y_i), that is:

$$Y_i = \alpha + \beta X_i, \ (\beta > 0).$$

A perfect relationship would demand that $\alpha = 0$ and $\beta = 1$. (See the diagonal line in Figure 3.10.) Theil proposes therefore an alternative inequality coefficient, Z:

$$Z = \frac{\sqrt{[(1/n)\Sigma(Y_i - X_i)^2]}}{\sqrt{[(1/n)\Sigma Y_i^2]} + \sqrt{[(1/n)\Sigma X_i^2]}}$$

where $X_1, X_2, \ldots, X_n$ are the levels in the first region and $Y_1, Y_2, \ldots, Y_n$ are the corresponding levels in the other. Except in the trivial case where all the X's and Y's are zeros, the coefficient Z is confined to the closed interval between zero and unity. $Z = 0$ in the case of perfect matching between the two series. A re-working of the Bristol sub-regional data using revised coefficients is currently being investigated.

Discussion

The results presented above suggest that small but detectable differences in the timing of cyclic movements in unemployment series exist between different geographical areas at different spatial levels. Such results are not inconsistent with the results obtained in a number of American studies (see the review by Bassett and Haggett, 1971, 390). In one of the more recent of these studies, King, Casetti and Jeffrey (1969, 214) test a model in which it is postulated that each city's activity level is affected only by (1) corresponding levels for other cities in the system and (2) exogenous national factors and their delayed consequences. The following system of difference equations provides a linear formulation of these linkages:

$$X_t = A_1 X_{t-1} + A_2 X_{t-2} + \ldots + A_n X_{t-n} + BY$$

where

$$X'_s = [x_{1s}, x_{2s}, \ldots, x_{Ns}] \ (s = t, t-1, t-2, \ldots, t-n)$$
$$Y' = [Y_t, Y_{t-1}, \ldots, Y_{t-n}]$$
$$A_s = [a_{sij}] \ (i, j = 1, 2, \ldots, N; s = t-1, t-2, \ldots, t-n)$$
$$B = [b_{is}] \ (i = 1, 2, \ldots, N; s = t, t-1, t-2, \ldots, t-n),$$

and where x_{it} is the economic activity level in city i at time t; y_t is the level of exogenous national factors which affect simultaneously all of the units

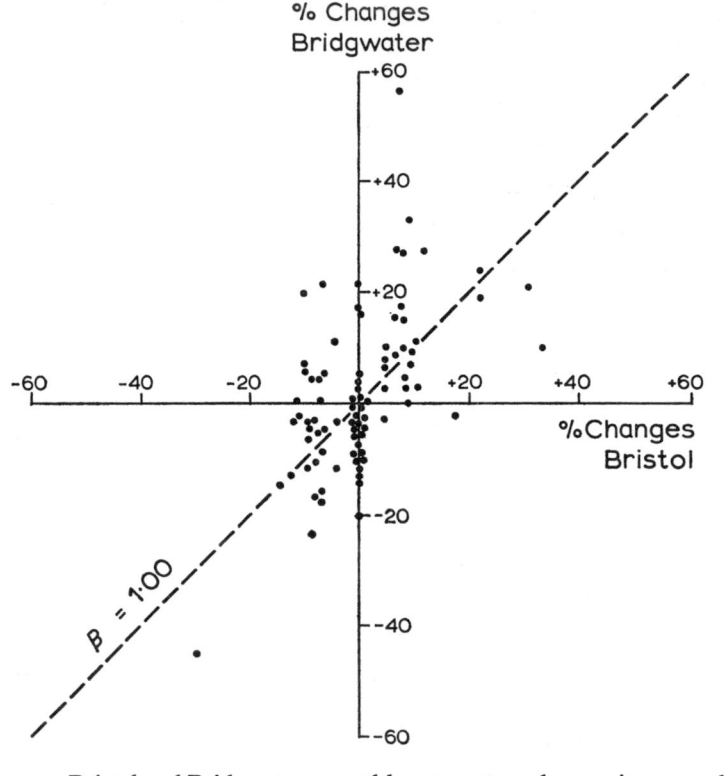

Figure 3.10. Bristol and Bridgwater: monthly percentage changes in unemployment levels in Bridgwater related to Bristol, 1960–9.

in the economic system at time t; a_{sij} is the influence on the ith city at time t of the economic activity levels in the jth city and time s; and b_{is} is the influence on the ith city at time t of the exogenous factors at time s.

In this system, considerable interest centres on the interpretation of the a and b coefficients. Zero a coefficients with strong non-zero b coefficients indicate strong exogenous effects on the cities within the regional economy with weak inter-city interdependencies. This situation might be characterised by reactions to across-the-board measures at the national level (e.g. bank rate changes, major legislation, devaluation, etc.). Conversely, non-zero a coefficients with zero b coefficients identify strong inter-city interactions with weak national linkages. In each case, we should expect the individual coefficients to vary, indicating to what degree, and after what time lag, cities are sensitive to changes in levels of economic activity in other cities and to exogenous national changes.

A partial testing of the model is provided by King *et al.* in a study of unemployment rates for thirty-three cities over twenty-six time periods. The data used were bi-monthly unemployment rates for May 1960 to September 1964 inclusive for midwestern cities (West Pennsylvania–Wisconsin area). Correlations between the national and city series were used as a surrogate for the *b* coefficients; lagged correlations between the residual city series (residuals from the regression of city series against national series) were used as a surrogate for the *a* coefficients. Results indicated three sub-systems of cities centred on Pittsburgh–Youngstown, Detroit and Indianapolis, with the two latter sub-systems lagging three to five months behind the first. Other cities (e.g. Chicago) had low levels of interaction with other cities but were strongly linked to the national system.

Perhaps the most common form of approach is to explore the lead–lag structure of different regions in terms of their industrial 'mix' or composition. In this approach, the business cycle is regarded as a national phenomenon; regional cycles reflect the national pattern in direct response to their share of national industries. A region where the industrial mix was in exactly similar proportions to the national industrial mix would, according to this hypothesis, be presumed to have a pattern of fluctuations exactly in accord with national fluctuations.

Such a relation between two regions may be shown as a simple linear transfer function. Figure 3.11 shows how a series of very simple relationships between a standard input wave (characteristic of lead region R_1) and an output wave (characteristic of lag region R_2) can be expressed as:

$$R_2' = \alpha + \beta R_1',$$

where R_1' and R_2' are differences in level between two successive time periods (e.g. per cent change in unemployment level over a month, as in Figure 3.10, and α and β are constants. When the output wave has exactly the same form as the input wave, $\alpha = 0$ and $\beta = 1$.

Changes in the α value result in regular change in the levels in the output region but changes in the value and magnitude of β cause change in the form of the wave. Values of β which are greater than $+1$ are deviation amplifying, while values of β less than $+1$ are deviation reducing (dampening). Negative values of β introduce inverse wave forms. It is clear that the examples given in Figure 3.11 are trivial and neglect expected non-linearities and feedback complications. However, more efficient estimates of wave transformation can be obtained for the South West and the hypothesis that industrial structures may 'amplify' certain wavebands but dampen or leave others unaffected may be tested.

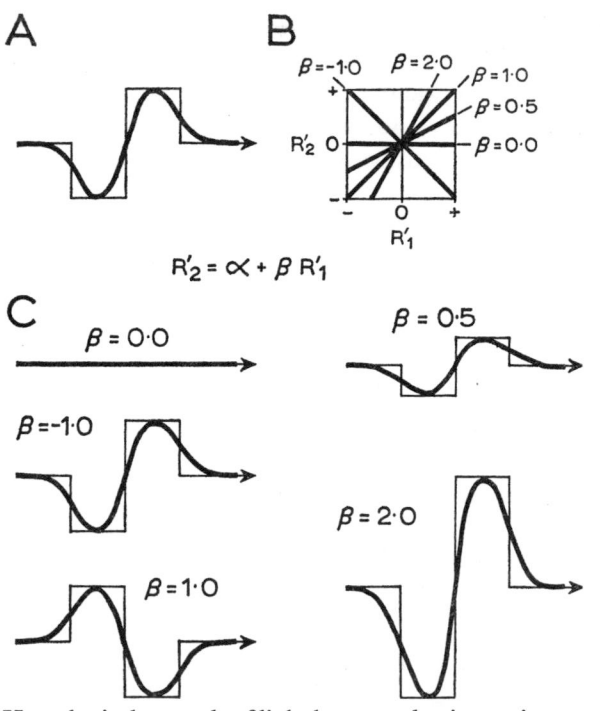

Figure 3.11. Hypothetical example of links between the time trajectory of unemployment values in one region (A) and other regions (C) expressed as a simple regression equation (B).

Note: positive values of β less than 1.00 suggest a 'dampening' effect while those greater than 1.00 suggest an 'accelerating' effect. In this idealised example, all feedbacks have been ignored.

It is difficult to see how the results on lead and lag structures in the Bristol sub-regions could be turned into direct forecasting use. Quite apart from the entrenched difficulties of autocorrelation and multicollinearity (Box, Jenkins and Bacon, 1967) there is a number of simple practical difficulties which make such an exercise hazardous. Figure 3.12 shows the results obtained by correlating series for the eight sample districts in the Bristol sub-region. It will be clear that whatever the precise form of the functional equations, no quarterly forecasts are possible for the three leading areas. Changes in these areas could be forecast only by reference to leading areas outside the region (perhaps in the West Midlands) or by a completely different type of analysis (perhaps using trends in indication other than unemployment). Second, accuracy of forecasts is directly related to the assumed likeness of a series in respect of the others. Thus, the

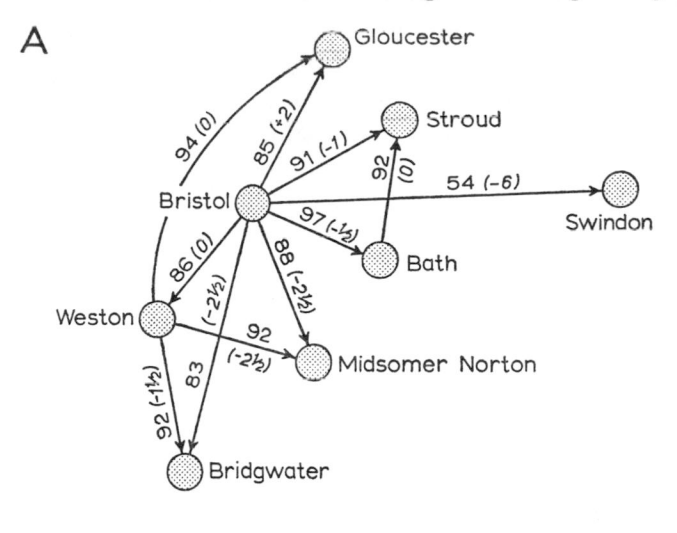

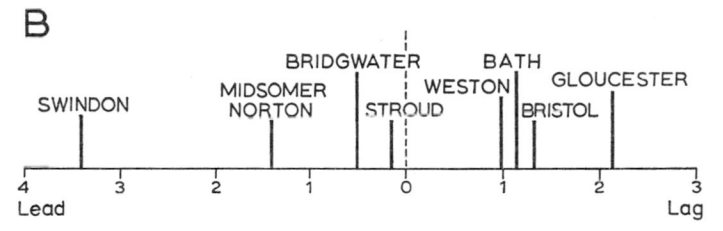

Figure 3.12. Bristol sub-region: maximum correlation between unemployment levels in districts in the sub-region (A) with average lead or lag relationship of each district to all others (B).

performance of Gloucester might be based on four other areas with which it had relatively high cross-correlations over the 1960-9 period (i.e. correlation coefficients of 0.81 to 0.93), while Stroud could be forecast only in terms of Swindon; and this with a correlation so low (0.21) as to indicate a likelihood of very poor forecasts indeed. Third, two of the leading areas are rather small employment areas (Stroud and Midsomer Norton) in relation to the larger areas being forecast. Small changes in either area are likely to cause rather large shifts in unemployment rates and introduce additional unreliability. Fourth, the forecasting models are based on statistical analysis over a fixed time period in the nineteen-sixties. It is possible that the lead–lag relationships obtained are wholly short-term phenomena related to the special character of industrial growth in the reference period.

For example, the strong lead shown by Swindon may be a short-term phenomenon heavily influenced by the first wave of Greater London Council overspill to the area and associated industrial investment. Although this sub-area is likely to maintain a significant growth rate in population in the next decade, the amount of industrial growth will depend on precise decisions on the phasing of development. The lag in the Gloucester area may be linked to the number of closures and redundancies in the last two years of the study period; these particularly affected the metal-using industries and firms with defence equipment contracts. The relative lag position of Bristol may also reflect short-term changes. During the nineteen-fifties and early nineteen-sixties, Bristol showed a rapid build-up in both the aircraft industry and in office industries. This growth wave petered out by the middle of the study period, with little change in manufacturing employment and only a small increase in service employment over the second half of the period. Such short-term changes point to the need for a 'rolling forward' of the calibration period of the model as new monthly data become available and the weighting of data in relation to their relative age, with more recent values being given higher significance (Bassett and Haggett, 1971, 412).

Finally, forecast models based on statistically derived lead–lag structures are not based on any logical or causal arguments. The models describe in a precise way the manner in which levels of unemployment move synchronously over space and time; they cast no light on the transmission mechanisms or the causal structure of the behaviour described. Certainly the lag structures described may reflect their relative degrees of employment size, diversification and sensitivity to national changes. But detailed studies of unemployment districts, like those already carried out by the South West Economic Planning Board for St Austell and Barnstaple–Bideford, will be needed before the precise significance of changes in local employment levels can be interpreted. Inter-regional transmission of change may also be seen more clearly when major input-output studies now being conducted for industry in the Bristol–Severnside area are complete (Edwards and Gordon, 1971).

If direct forecasting proves infeasible then it is conceivable that lead–lag information might prove useful in simulating the dynamic properties of regional systems over time. For example, it is possible to conceive unemployment levels in a set of regions in system terms, so that changes in levels over time are linked through lead–lag structures to levels in other regions (Di Stefano, Stubberud and Williams, 1967). A very simple but illustrative case is presented by Blalock (1969, 79) who describes a difference equation model for lagged endogenous variables that can be adapted to a simple

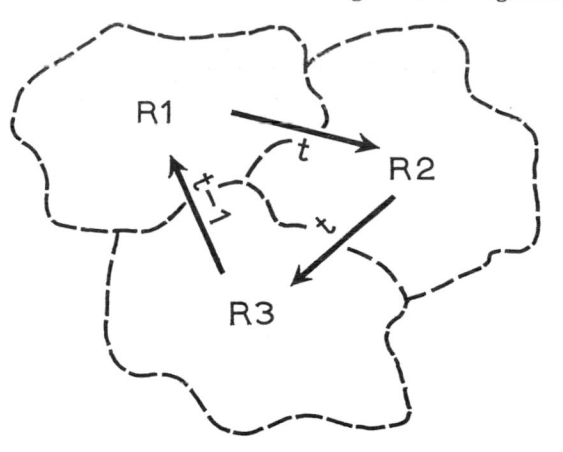

Figure 3.13. Simple three-region system with lag relationships.

three-region case. Figure 3.13 gives a simple explanation of inter-dependencies between regions in the form of a closed loop with arrows running clockwise and ending at the starting point. However, we specify that the links between R_1 and R_2 and between R_2 and R_3 are instantaneous, but that the links between R_3 and R_1 are delayed. Such delays have been shown by King, Casetti and Jeffrey (1969, 218) as operating between systems of cities in the American mid-West, e.g. the Detroit complex lags three to five months behind the Pittsburgh complex. Our three-region case is highly simplified but it is small enough to handle effectively in a simple set of difference equations.

Let the level of economic activity in region R_1 at time t be written as R_{1t}, the level of economic activity in R_3 at a preceding time period as $R_{1,t-1}$, and so on; then we can formulate the system shown in Figure 3.13 as:

$$R_{1t} = a_1 + b_{13}R_{3,t-1} + U_{1t}$$
$$R_{2t} = a_2 + b_{21}R_{1t} + U_{2t}$$
$$R_{3t} = a_3 + b_{32}R_{2t} + U_{3t}.$$

In each case, the U terms refer to uncertainties in the estimate or error terms. Blalock goes on to show how substitution in these equations can yield:

$$R_{1t} = A_1 + b_{13}\,b_{32}\,b_{21}\,R_{1,t-1} + (U_{1t} + U_{2,t-1} + U_{3,t-1}).$$

In other words, we can now express the level of economic activity in R_1 at time t as a function of itself at time $t-1$ plus various error terms. Inspection of Figure 3.13 shows that we have simply completed the circular feedback loop shown by the arrows.

TABLE 3.1. *Types of feedback situation for the difference equation model* $X_{1t} = A + b_{13}b_{32}b_{21}X_1$, $t-1$ *applied to a three-region system*

		Feedback situation	
		Positive	Negative
Stability situation	Stable	Type II $b_{13}b_{32}b_{21} > 0$ $\lvert b_{13}b_{32}b_{21} \rvert < 1$	Type IV $b_{13}b_{32}b_{21} < 0$ $\lvert b_{13}b_{32}b_{21} \rvert < 1$
	Unstable	Type I $b_{13}b_{32}b_{21} > 0$ $\lvert b_{13}b_{32}b_{21} \rvert > 1$	Type III $b_{13}b_{32}b_{21} < 0$ $\lvert b_{13}b_{32}b_{21} \rvert > 1$

The b coefficient in the equation is an indicator of what might happen to levels of economic activity over time, if we ignore the effect of the disturbance terms. If the numerical value of the product of $b_{13}b_{32}b_{21}$ is less than unity, successive changes in the level of R_1 will die away; conversely, if they are greater than unity each successive time interval will bring further increases. Types of feedback situations are identified in Table 3.1 and typical regional trajectories are suggested in Figure 3.14.

Of course, there is a number of compelling reasons why the simple three-region model is inappropriate. From the purely statistical point of view the estimation of the parameters poses serious difficulties and runs into the classic econometric problems of autocorrelation bias. From the practical point of view, large numbers of regions are involved with two-way interdependencies with complexly distributed lags which may or may not be stable over time. The point being sought here is not to suggest that a difference equation model is appropriate, but that it points a way towards writing models of regional interconnections in terms of systems of differential equations. The advantages of such systems are that they focus attention on critical parameters and narrow the search area for significant attributes of regional interconnections.

Some evidence for negative forms of feedback in regional unemployment levels is presented in Brechling's (1967, 17–18) study of regional cyclic components for ten British regions. Two distinct cyclical patterns were isolated. The South West region forms a group with London, the South and East, Wales, the North Midlands and the Midlands with lows in 1954–5 and 1959–60 and highs in 1957–8 and again in 1961–2. The remaining

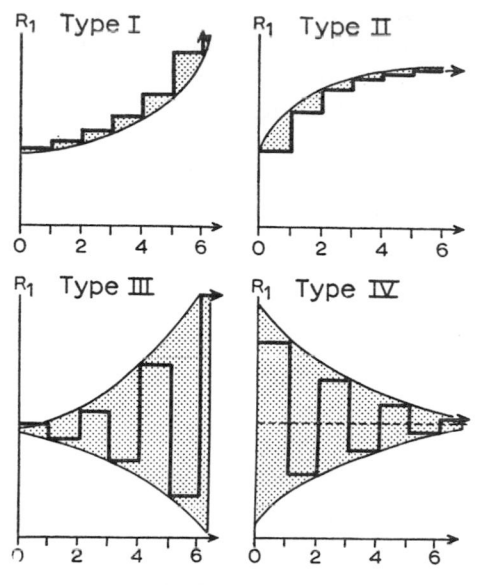

Figure 3.14. Alternative paths of feedback situation and their effect upon regional trajectories.
Note: see Table 3.1 for identification of the four types.
Source: Blalock, 1969, 82.

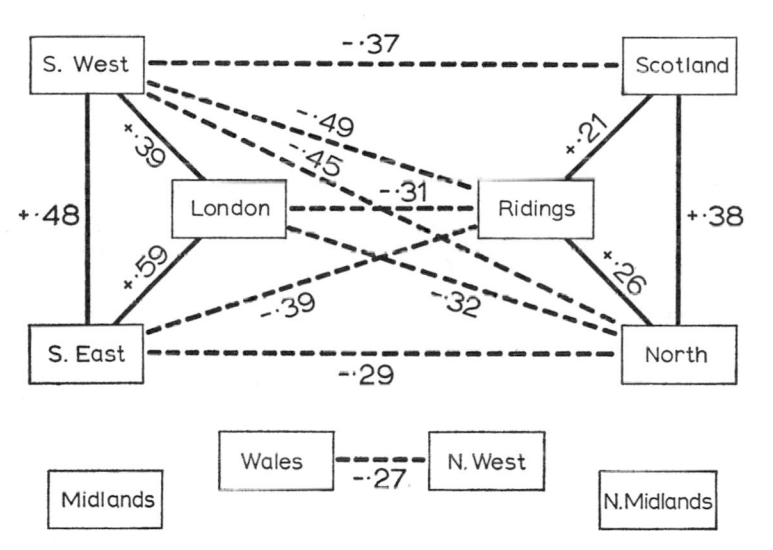

Figure 3.15. Great Britain: assumed feedback relations between unemployment levels in major regions as shown by correlations between residuals.
Source: Brechling, 1967, 18.

regions (the Ridings, the North West, the North and Scotland) showed an inverse pattern with highs when the other group showed lows and vice versa. Figure 3.15 shows the detailed form of this pattern from Brechling's zero-order correlation of the residuals. Only R^2s of greater than 0.20 are shown and negative bonds are indicated by broken lines.

Recognition of cyclic patterns and the identification of lead–lag structures represents only a preliminary stage in the description of regional hierarchies in system terms. The long-run importance of this formulation is likely to be seen in terms of public policies for system control. Forrester (1969, 9) argues that 'complex systems are counter-intuitive; they give indications that suggest corrective action which will often be ineffective or even adverse ...policies adopted for correcting a difficulty [may] actually intensify it rather than produce a solution'. Much of this difficulty seems to stem directly from the linking of regions in terms of feedback loops and the delicate nature of the boundary between positive (system amplifying) and negative (system damping) loops.

Conclusion

This chapter has presented a series of studies which show that for one indicator of economic health (the level of unemployment), small but detectable differences in the timing of cyclic movements may be demonstrated at a series of spatial levels within the South West region. For the most part, such leads and lags are small in magnitude and their importance may be swamped in many parts of the region by the much greater amplitude of seasonal swings or irregular components. The main contribution of such studies would seem to be in (1) the calibration of cyclic elements as short-term (quarterly) forecasting models for levels of sub-regional activity and (2) the development of simulation models in which the effects of the introduction of internal and external 'shocks' might be recorded for a regional system. The long-run importance of work of this kind for policy issues as opposed to academic model-building is more debatable. It is arguable that, in the very long run, disaggregated models might be built in which the regional effects of alternative central-government policies might be tested and re-framed within appropriate simulations. The results reported here suggest that an awareness of leads and lags or phase differences between the reactions of various parts of the regional system in different geographical areas and at different spatial scales may play some very small part in the design of such models.

Acknowledgement

The study reported in this chapter stems from work in the Department of Geography, University of Bristol, undertaken jointly with Mr K. Bassett and Dr A. D. Cliff under a grant made by the Social Science Research Council.

References

Bassett, K. A. and P. Haggett (1971). 'Towards short-term forecasting for cyclic behaviour in a regional system of cities', in M. Chisholm, A. E. Frey and P. Haggett (eds.), *Regional Forecasting*, Butterworths.

Bassett, K. and R. Tinline (1970). 'Cross-spectral analysis of time series and geographical research'. *Area*, 1, 19–24.

Blalock, H. M. Jr. (1969). *Theory Construction*, Prentice-Hall.

Box, G. E., G. M. Jenkins and D. W. Bacon (1967). 'Models for forecasting seasonal and non-seasonal time series' in B. Harris (ed.), *Advanced Seminar on Spectral Analysis of Time Series*, Wiley.

Brechling, F. (1967). 'Trends and cycles in British regional unemployment'. *Oxford Economic Papers*, New Series 19, 1–21.

Bronfenbrenner, M. (1969). *Is the Business Cycle Obsolete?* Wiley.

Central Statistical Office (1969). *Economic Trends*, 185, HMSO.

Di Stefano, J. J., A. R. Stubberud and I. J. Williams (1967). *Theory and Problems of Feedback and Control Systems*, Schaum.

Edwards, S. L. and I. R. Gordon (1971). 'The application of input-output methods to regional forecasting'. In M. D. I. Chisholm, A. E. Frey and P. Haggett (eds.). *Regional Forecasting*, Butterworths.

Eversley, D. (1968). *New Society*, January 4, 1968.

Forrester, J. W. (1969). *Urban Dynamics*, MIT Press.

Isard, W. (1960). *Methods of Regional Analysis*, MIT Press.

Granger, C. W. J. (1964). *Spectral Analysis of Economic Time Series*, Princeton University Press.

King, L. J., E. Casetti and D. Jeffrey (1969). 'Economic impulses in a regional system of cities'. *Regional Studies*, 3, 213–18.

Prest, A. R. (ed.) (1968). *The U.K. Economy*, Weidenfeld and Nicolson.

Reddaway, W. B. (1970). *Effects of the Selective Employment Tax: First Report on the Distributive Trades*, HMSO.

South West Economic Planning Council (1967). *Region with a Future*, HMSO.

Theil, H. (1958). *Economic Forecasts and Policy*, North-Holland Publishing Company.

Weltman, J. and E. Rendel (1698). 'Unemployment indices'. *Quarterly Bulletin of the Research and Intelligence Unit, Greater London Council*, 3, 21–5.

4. Spatial structure of metropolitan England and Wales

PETER HALL

This chapter has two objectives. The first and more general is to discuss the conceptual and technical problems of defining urban areas. (The term 'urban areas' is used in preference to particular words like 'towns' or 'cities', because it is the most general expression, without specific conventional connotations.) This is done in the first section. The second and more particular is to consider the applicability to Britain of one well-known definition of urban area: the Standard Metropolitan Statistical Area. This is done in the second section. The third section presents some results of the exercise; and a discussion of some policy conclusions rounds off the chapter.

The general problem

Definitions

That there is a problem and that it seems important, is evident from the literature. Statements are frequently made there about the percentage of the population of different nations that is urban, or about the growth of urban population in different nations as a percentage of their total population growth; but on inspection, such comparisons prove to depend on quite different national definitions of what is urban. In Denmark a place with 250 people is urban, in Korea a place with less than 40,000 is not (Hall, 1966, 19). So there is need for international standardisation. And even within any one nation, there is need for a closer definition of what is meant by an urban area. The definition used by the British Census, for instance, depends wholly on the administrative designation of an area as urban or rural, a designation which may reflect past rather than present reality; it is at least odd that between 1951 and 1961 and between 1961 and 1966, the rural areas were increasing in population faster than the urban areas (Hall *et al.* 1972, ch. 5). The problem, therefore, seems *prima facie* to be worth discussing. To the layman, and perhaps to anyone up to the present century, it might seem a problem with a simple solution. The simple answer to the question 'What is an urban area?' might be, 'An area that looks like town'. A slightly less simple answer might be an 'An area that works like a town'. These are, indeed, the basic clues to the answer. At first they may not seem to present difficulties. Laymen might think that they could recognise a town if they saw one, as they could identify a townsman by asking him a few questions. Through most of historic time, in fact, this might have sufficed. There was a sharp physical break between town and country, as an old Italian or Dutch painting will show. Townspeople did different jobs from country folk, and lived in different sorts of societies. On either count, the physical or the functional, there was no difficulty in separating town from country, and the two definitions produced identical results.

But that simple state of affairs no longer obtains. To some extent even in Britain, but far more so in North America or in Australia, towns spread physically into the countryside in the form of suburbs, at decreasing densities as one goes farther out. Colin Clark has shown that this phenomenon progressively extends over time; so urban and rural population densities, once so clearly differentiated, fade into each other along a continuum (Clark, 1951; 1967, Chapter 8), and their physical manifestations in bricks and mortar do the same. Many outer American suburbs have lower population densities than many areas of South East Asia, where the great majority of the population depend on agriculture, and yet the latter should be classed as rural in some functional sense. Again, in most advanced industrial countries it has been observed that many city workers choose to live in the countryside; a farmhouse, or old mill, may be occupied by a family whose breadwinner depends on a job done in the centre of a city. As the length of the working week declines, this pattern becomes easier to pursue; and the development of electronic communications may entirely free the 'urban' dweller to live where he wishes in the countryside (Whyte, 1970). There is no longer any simple accord between the physical and the functional ideas of urbanisation.

Having stated the problem, let us approach a solution by trying to define, more closely, what we mean by these two possible meanings of an urban area: the physical, and the functional.

The physical meaning obviously conveys the idea of a congregation, or concentration, of buildings. A town, or city, is commonly thought of as being a bigger, or denser, concentration of buildings than a village, or hamlet. But sometimes this rule breaks down. There are so-called villages in parts of southern Germany which have more buildings, just as densely concentrated, as some so-called market towns in the same area; they are called villages because most of their inhabitants used to perform agricultural functions. And if these people turn from farming to commuting into factories, as has happened to a large extent since 1950, then presumably the village turns into a town, without any change of physical structures. So for the sake of completeness, there are two necessary dimensions to the physical meaning. One is a definition of a certain sort of building: an *urban* building. In fact this may well be a functional and not a physical definition; for as we just saw, houses may change their purposes. But often, because it is thought to be self-evident whether the buildings are urban or rural, or because of simple lack of statistical data on the subject, this dimension is omitted. Perhaps indeed some proxy for the physical form may be employed: thus instead of houses, the people who live within them may be counted. The other dimension is intensity of occupation of the ground, or density: for instance dwellings, or habitable rooms (or as we just saw, people used as a proxy), per hectare or per acre.

Physical meanings of urban areas are commonly employed in official statistics. But often, one or even both of the dimensions are left unmeasured. The large conurbations, which have been used as a basic spatial unit in British Censuses since 1951, are physical urban areas, defined in terms of continuous built-up terrain (General Register Office, 1956, xv). But no systematic attempt is made to define what sort of built-up area is involved, or what intensity of building qualifies for inclusion. The urbanised areas, which have been used as a basic definition in the United States Censuses since 1950, are defined in terms of minimal density of population, not of physical structures; and no attempt is made to define what sort of population should qualify for consideration. Furthermore, the minimal density used – 386 persons per square kilometre (1,000 per square mile) or 3.75 per hectare (just over 1.5 per acre) – is frankly arbitrary. And in practice, it proves difficult to define a cut-off point that is not.

The alternative, functional, meaning obviously conveys the idea of an area containing activities, which have been defined as urban. There is in fact an element of circular reasoning here: activities which have been defined as urban must be activities which have been associated with places which have been physically defined as urban, even though they might not necessarily be associated with such places always and in all places. The United States Census definition of a Standard Metropolitan Statistical Area, perhaps the best-known example of the official use of a functionally defined urban area, includes a criterion of metropolitan character; to be included in an SMSA, a county must have at least 75 per cent of the total labour force counted as non-agricultural labour, as well as certain other characteristics (Bureau of the Budget, 1964, 1–2). This automatically defines the urban area, functionally, in terms of a certain range of occupations, which traditionally have been regarded as urban. But in fact, it is not self-evident which occupations should be regarded as urban, and which not. It is at least arguable for instance that most extractive industry (mining and quarrying) is not urban in character, any more than reproductive industry (agriculture, forestry and fishing) is. Yet the United States Census definition automatically includes the size of the labour force in extractive industry as contributing to the determination of metropolitan character.

The nature of the activity, or activities, is one dimension of the functional definition of urban area. Such activities may be static, or within-place, or they may be dynamic, or between-place. An example of a static (within-place) activity, used in this sense, is the one we have given: the percentage of the labour force engaged in non-agricultural occupations. An example of a dynamic (between-place) activity may also be taken from the United States SMSA; it is one of the principal criteria of metropolitan integration. There, a county is regarded as integrated with the county or counties containing the central cities of the SMSA if either 15 per cent of the workers living in the county work in the county or counties containing central cities of the SMSA, or if 25 per cent of those working in the county live in the county or counties containing central cities of the area (Bureau of the Budget, 1964, 2). Here we are dealing with flows, or interconnections, or interrelationships, between one place and another, at least one of which has been defined as urban on some other ground. Such interrelationships might include shopping patterns, educational patterns, telephone or letter messages, recreational patterns, water or milk supply patterns, or a great variety of other phenomena. In each of these examples, one end of the interchange would previously be defined as urban (the shopping centre, the destination of the milk, the origin of the residential trips); the object would be to define the sphere of influence of that end of the interaction. As with static activities, so with dynamic ones, there is room for much disagreement about which are urban relationships, and which are not. Most workers have tended to allow that commuting patterns, and shopping patterns, are permissible bases for defining functional urban areas; not all might agree that water or milk supply areas were allowable.

The nature of the activity, or activities, is one dimension of the functional definition of an urban area. The other of course is the intensity of the activity. We may allow that a certain density of resident population, or daytime working population, defines an urban area. Or we may define that area in terms of a certain intensity of commuting, or shopping, or some other kind of trip. As with definitions, it may prove hard to find a cut-off point that is not arbitrary. To try to overcome this, we should try to array and inspect our data, to see whether they show significant break-points. For instance, we should plot intensity of commuting by distance from origin to destination, and see whether there are strongly marked residuals from the general regression line; but unfortunately, in many cases the fit of the data to the function is extremely close.

If there are interesting irregularities, we may try to interpret them by disaggregation. Commuting patterns, for instance, may often best be interpreted in terms of a strongly exponential function, in which the graphical plot shows a long tail of long-distance commuters – a minority, who behave very differently from the great majority (Royal Commission on Local Government, 1968, 444). We may well suspect that this minority is different in many ways – in income, in socio-economic group, in car owner-ship, in educational history, in the amount of information possessed about alternative opportunities – from the majority. The point is that when we try to define a functional urban area, we need to ask 'whose urban area?'. The functional area of the rich is not the same as that of the poor, nor is the area of the old that of the young, nor of men the same as that of women. Within the same family, the functional area of the male breadwinner is by no means the same as that of his wife or small children, as Swedish work on household time and space patterns is just beginning to confirm (Häger-strand, 1970). These differences in pattern will emerge for almost any type of interrelationship. But the pattern is further complicated by the fact that every individual and every group has different sorts of interrelationship, some of which show much greater richness of variation than others. Most rational people do not move far when they buy a newspaper or take their dog for a walk, whether they are old or young, rich or poor. But apparently rational people show a much greater variation in their commuting patterns, and a greater variation still in their holiday-making patterns. Ever since the work of Christaller, geographers have recognised that higher-order goods and services will attract people over longer distances than lower-order ones (Christaller, 1936; 1966). The range of possible variation is greater for the higher-order relationships than for the lower.

Bearing this in mind, it can be accepted that there is a need for disaggre-gating each pattern of movement. Recent work on urban modelling has suggested this need very strongly; commuting patterns for instance need to be analysed by car ownership and possibly also by income (which are strongly correlated with each other); residential location patterns (and therefore, implicitly, residential migration patterns) should be disaggregated by type of house and household income (Wilson, 1969, 46–53). If data are available, it will be interesting and perhaps fruitful to disaggregate the relationships in as many different ways as possible; and then to simplify the results by using some form of logical grouping procedure, such as factor analysis. But even if such re-aggregation is achieved, it will not avoid the problem of arriving at a single working definition of a functional area. This is merely an illustration of the well-known general problem of deriving composite indices, and it must involve selection and weighting procedures. If complete data were available on all movements made by all members of households (or, of course, of a carefully-taken sample of households), then it should be possible to develop a weighting based on the relative importance of different movements. We could for instance define the functional urban area as that which encompassed 75 per cent (or 85 per cent, or any desired proportion) of all recorded interactions. Or we could try to find significant break-points in the data. But despite the work of the Swedish geographers, such data are not generally available.

Data

We are in fact very dependent, as usual, on the limitations of the data that do exist. In most advanced industrial countries, such as Britain, four principal data sources can be distinguished. The first is the regular Census count of journey-to-work movements. Commonly this is published (or is available unpublished) in the form of a matrix of origins and destinations. Usually these refer to conventional administrative–statistical units such as local government areas, which may vary greatly in size; there is conventionally a minimum cut-off in terms of intensity of interaction, below which the data are not published. These features present considerable difficulties in interpreting the data, particularly if the information is based on a small sample of the entire population, as in the British Censuses of 1961 and 1966. But against this, the Censuses have the merit of regularity, uniform coverage and general consistency (with some variations in detail) over a long period. Data in broadly comparable form, covering the whole country, are available for instance from the British Censuses of 1921, 1951, 1961 and 1966.

The second source is *ad hoc* surveys of shopping or other movements. Invariably these are one-shop surveys, made at a single point in time, usually for one area only (for instance Kent County Council, 1963). They are not easily comparable with each other, and provide scant basis for analysis between regions or over time periods.

The third potential source, which is growing in importance, is the land use–transportation studies made for major urban areas as part of the transportation planning process. In the United States and Britain (but to a lesser extent, at the time of writing, in Continental Europe) such studies already cover a major part of the total population and economic activity and therefore a major part of the total potential interaction in the whole country. They have the merit of being presented in reasonably standardised form (in terms of the classification of trip purpose, and in terms of the period to which the survey refers) for small zones, designed specifically to be reasonably similar in terms of population or economic activity. Yet there are difficulties: the inventories are of a one-shot type, taken at different times in different urban areas, and they are not usually repeated, since updating is accomplished more simply by sample checks; the samples involved are very small; the external cordon line, which bounds the area for which detailed zone-by-zone information is given, is sometimes very narrowly defined; and by no means all types of movement are included, since walking trips (for instance) are commonly excluded from many of the inventories. In combination, these difficulties make it virtually impossible to use the transportation study inventories for any significant comparison between different areas or between different dates.

Finally, there are some *ad hoc* surveys which show patterns of inter-communication: goods vehicle tons moved, letters or telephone calls conveyed, and similar measures. They are invariably national surveys which present their data on rather coarse geographical bases, which do not relate easily to conventional areas, and are hence not comparable from one survey to another; so they are difficult to build into an integrated picture.

It is not surprising, therefore, that in practice urban analysts have made disproportionately great use of Census journey-to-work data in defining functional areas. These data have the additional advantage that they can be directly related to other Census data for static phenomena like residential population (total, or employed) and workforce, which are available for the same years on the same geographical base. They may even be associated with other Census data, on phenomena like housing, which may be used to produce a physical definition of urban area directly related to the functional one. Lastly, they do refer to the most common single type of regular movement made in urban areas. As so often, therefore, there is inevitably a great difference between the ambitious theoretical analysis that might be possible if data were available, and the relatively mundane and simple analysis that has to be based on the available facts. The next part of this chapter will accordingly deal with the possible application to British urban analysis of perhaps the best-known functional definition of an urban area, which happens to be based in large measure on commuting data: the Standard Metropolitan Statistical Area of the United States Census.

Standard Metropolitan areas

The British journey-to-work data are among the richest in the world: in their range and completeness over time, they are far superior to those of the United States Census. Yet with a few exceptions, such as Kate Liepmann's pioneer study of journey-to-work (Liepmann, 1944) and some valuable analysis by R. Lawton (Lawton, 1959; 1963), most British work on functional urban delimitation has been based on service hinterlands rather than on commuting (Smailes, 1944; Green, 1950; Carruthers, 1962; 1967). The British Census, moreover, has produced no official definition of a functional urban area based on commuting data. Yet in the United States, such a definition has been a feature of every Census since 1940. Known then as the Metropolitan Region, this area became known in 1950 as the Standard Metropolitan Area, and in 1960 as the Standard Metropolitan Statistical Area (SMSA). It is now used not only by the Census but by many other official statistical agencies, by academic researchers and by planners. It is significant that the most ambitious work on functional urban delimitation yet accomplished in Britain – part of a world-wide comparative study of the subject – was carried out by an American agency, and tried to apply the Standard Metropolitan Area concept (Davis, 1959, 1–33). Because of the availability of data, therefore, it seems useful to consider again the specific problems of applying such a concept to the British scene.

The Standard Metropolitan Statistical Area, as used in the 1960 United States Census, is built up of counties (except in New England, where townships are used instead); in the eastern United States these are invariably smaller than their English equivalents, being intermediate between an English county and an English county district, though in the western part of the country they tend to be much larger, the extreme case being over 320 kilometres (200 miles) across. The definition of the SMSA then involves three separate stages.

First, there must be a city of a specified population – in 1960 50,000 – constituting a central county; in some cases, a number of counties may together constitute the central core of the SMSA. It will be seen that the idea of centrality is essential to the SMSA concept; a completely dispersed urban area, as seems to be evolving in areas of the western United States like Phoenix (Arizona) or Las Vegas (Nevada), (Riley, 1967), would be increasingly difficult to define using the SMSA procedure.

Second, contiguous counties are investigated to see whether they fulfil the condition of metropolitan character. As stated earlier, the 1960 definition of metropolitan character was essentially that 75 per cent of the labourforce must be non-agricultural and that it must live in contiguous minor civil divisions with a population density of at least 58 persons per square kilometre (150 per square mile).

Third, and additionally, these counties are investigated to see whether they fulfil the condition of metropolitan integration. The criteria used in 1960 were that 15 per cent or more of the resident workers in each of these counties should commute to the county or counties containing the central city, or that 25 per cent of those working in the county should live in the county or counties containing central cities of the area. Additional criteria of integration could be considered if data for these criteria were not conclusive (Bureau of the Budget, 1964, 1–3).

From this account of procedure, it can be seen that the SMSA is a functionally defined urban area based on both static criteria (those of metropolitan character) and on dynamic criteria (those of metropolitan integration). In general, but with particular respect to the latter, it is clear that the operation of the criteria will depend a great deal on the size of the building blocks used. Small building blocks (eastern US counties) may produce a completely different picture of commuting, for instance, as compared with large ones (western US counties). The size range of English administrative areas, for which commuting and other data are published in the Census, is certainly smaller than the size range of American counties. But the problem needs to be kept in mind.

In terms of the general discussion in the first section of this chapter, it is clear that the Standard Metropolitan Statistical Area is an exceedingly narrow type of functional urban area definition. It deals basically with one activity and one interrelationship: work, and the commuting patterns which the geography of work and of residence produce. Consequently, it deals not with the entire activity patterns of whole families, but with one aspect only of the activity patterns of those members of families who happen to be in employment. (There is an obvious justification for this choice: that for many adults work is the most time-consuming human activity apart from sleep, and the journeys are the most common single form of journey, as revealed by all land use–transportation studies; and that work generates incomes which, brought home by the commuter, generate further income and employment in the residential area.) Nevertheless, the narrowness and exclusiveness of the choice must be admitted. Equally obvious is the simple assumption that there is a definable central place to the commuting system, and that a meaningful level of interaction can be defined in terms of a certain intensity of commuting.

The SMSA concept can be criticised on virtually every one of its basic criteria of compilation, and has been so criticised by Berry and his colleagues in a study for the United States Social Science Research Council (Berry, 1967). They argued that the definition of the central city in terms of legal limits was arbitrary; that the minimum population criterion for the central city was similarly arbitrary; that the criteria of metropolitan character were irrelevant, since the great majority of American society were urbanised wherever they lived; and that the 15 per cent cut-off level for commuting to the city was also arbitrary. All these criticisms would seem to have equal force for any application of the concept to British experience – some of them even more force. For instance, the non-agricultural element in the British labourforce, even in the countryside, is normally much higher than 75 per cent of the total, and very few English rural districts near large or medium cities do not reach the minimum population density of 58 persons per hectare (150 per square mile). The criterion of metropolitan character, therefore, could probably be discarded altogether for British purposes.

The problem of central city definition is more complex. Even in the case of small and medium-sized freestanding towns, cases occur where the town is enlarged to swallow up completely a former urban district; in this case the so-called central city may become larger, without any real functional change having occurred. And in the case of the larger, more complex urban agglomerations it becomes very difficult to define precisely what is meant by the term central city. Thus in making a pioneer attempt to apply the American SMSA concept to England, Kingsley Davis and his colleagues at Berkeley (California) took the central city of the London metropolitan area to be the old London County Council area, while in the West Midlands they took as basis the whole of the West Midlands conurbation, an apparently much more generous definition (Schnore, 1962). It seems desirable therefore to find an objective criterion that is, as far as possible, independent of administrative details. If the central city is to be regarded as the chief single centre of the urban area for employment, and hence for in-commuting, as in England it invariably still is, then logically it should be defined in terms of levels of employment; and the best single comparative index that could be devised would be the density of employment, per hectare or square kilometre. Unfortunately, it proves very difficult to apply this successfully in all medium-sized freestanding towns, some of which have very generous boundaries with consequent effect on their employment density. Among the greatest English cities, for instance, employment density ranged from 38.8 per hectare (15.7 per acre) in Manchester (1961) to 16.6 per hectare (6.7 per acre) in Leeds: so 12.4 workers per hectare (5 per acre) would seem a reasonable cut-off value.

But among smaller cities, the density fell as low as 7.7 per hectare (3.1 per acre) in Southport and 8.2 per hectare (3.3 per acre) in Shrewsbury. It is desirable therefore to allow an alternative criterion for such small free-standing towns, in the form of an absolute minimum level of employment of (say) 20,000. But it should be self-evident that some minimum cut-off level is desirable, otherwise, logically, one would be forced to consider a very large number of very small commuting systems. A close look at the actual commuting flows in any part of the country shows a pattern often dominated by very short-distance local flows, to centres that are often quite small. The minimum cut-off to be chosen, therefore, depends on the purpose and the scale of the analysis: if this is to be local in character it may be quite low, but for a national comparison it should be quite high. The application of the two alternative criteria (12.4 workers per hectare or a total of 20,000 workers) above gives exactly 100 centres of metropolitan areas in 1961, which is a convenient enough figure for national analysis. The United States Census of 1960, in comparison, had 219.

The next problem is the criterion to be used for integration. As argued above, the limitations of data make it almost inevitable that journey-to-work information will be the basic source, in Britain as in the United States. So the critical question again concerns the minimum cut-off level. In Britain, the Berkeley researchers apparently used the figure they had inherited from the United States Census: 15 per cent (Davis 1959, 27). Berry and his co-workers, by detailed analysis of Census tracts, were able to show that often this level is arbitrary and too narrowly exclusive (Berry, 1967, 2–3). This last point has less force in England, where even at the 15 per cent level the commuting fields exhaust a very large part of the central industrialised belt of the country; extending the boundaries out to the farthest limits of the commuting field, conversely, will produce a very considerable degree of over-lap, and if smaller centres are admitted to the analysis this will prove even more apparent. At any rate, it seems desirable to recognise two alternative cut-off points: one the traditional 15 per cent level of the United States Census; the other, Berry's suggested Metropolitan Economic Areas, which like the SMSA's are based on central cities of 50,000 and more people, but which extend to take in all the areas sending commuters into that city (except those areas sending more commuters to another central city).

These in fact were the two alternative definitions used in the study of urban growth in Britain since 1945, carried out by a team at Political and Economic Planning (PEP) in London, in association with a parallel study in the United States at Resources for the Future Inc. in Washington, D.C.

(Clawson, 1971). An important reason for choosing both definitions, and for the terminology used, was the need for close comparability with the units employed in the American study. To be precise, the units were:

(i) *The Standard Metropolitan Labour Area (SMLA)*. The word 'Labour' is added to indicate the importance of employment for defining the central core.

The SMLA consists of:

An SMLA core, consisting of an administrative area or a number of contiguous areas with a density of 12.5 workers and over per hectare; or a single administrative area with 20,000 and more workers.

An SMLA ring, consisting of administrative areas sending 15 per cent of their resident employed populations to the core, and forming a group contiguous with that core. To be regarded as an SMLA, the whole group should have an enumerated population of 70,000 or more.

(ii) *The Metropolitan Economic Labour Area (MELA)* which consists of:

An SMLA core ⎫
An SMLA ring ⎭ identical to the above.

The outer MELA includes all administrative areas not included in the SMLA core or ring, but forming a group contiguous with either, in which each area sends more of its resident employed population to the SMLA core than to any other SMLA core. Any area is included here which sends any commuters to the SMLA core, provided it does not send more commuters to any other SMLA core and that it forms part of a contiguous group touching the SMLA core or ring.

The SMLA therefore fits inside the MELA (except in those few cases where the two are identical) like one box fitting inside another.

A critical problem for the definition of all such areas, if they are to be maintained and compared over time, is whether the criteria are to be re-applied at each successive Census date, thus producing new definitions. The United States Census allows the definition of its SMSAs to float in this way, but the British study team decided against it. The advantage of a floating definition is that it truly reflects the functional area at that moment according to the criteria employed; the disadvantage is that exact comparisons between one date and another are impossible. Because the time span used for the British study was effectively short, 1951–66, it was decided that the balance of advantage lay with the fixed definition. The limiting values were therefore taken from the 1961 census.

Some results

Taking first the more limited SMLA definition, out of some 46.10 million people counted in the Census of England and Wales in 1961, no less than 35.72 million (77.5 per cent) lived in the 100 SMLAs. This proportion had hardly changed in the previous thirty years; it had been 77.0 per cent in 1931, 77.3 per cent in 1951 and it was to fall slightly to 77.1 per cent in 1966. The metropolitan areas were thus adding population about as rapidly, in percentage terms, as the nation as a whole; they added just over three million between 1931 and 1951, close on 1.9 million between 1951 and 1961, but only 640,000 between 1961 and 1966. The corresponding decennial rates of growth were 5.0 per cent from 1931 to 1951, 5.6 per cent from 1951 to 1961, but only 3.6 per cent from 1961 to 1966.

Extending now the analysis to the more generously defined MELAs, we find that 100 Metropolitan Economic Labour Areas, based on exactly the same central cores as the SMLAs, contained 43.29 million people (93.9 per cent of the total for England and Wales) at the 1961 Census. This proportion also had remained almost constant: it had been 93.3 per cent in 1931, 93.7 per cent in 1951 and was 93.9 per cent in 1966, the same as in 1961. Like the SMLAs, the MELAs had a faster growth in percentage terms from 1951 to 1961 than from 1931 to 1951; but from 1961 to 1966, their rate of growth slowed down.

The concentration of employment was slightly more marked than that of population, whichever of the two units is used. In 1951 (no employment figures by workplace are available for 1931) the SMLAs contained 79.0 per cent of the total employment in England and Wales; the proportion rose to 79.4 per cent in 1961 and to 79.5 per cent in 1966. The corresponding figures for the MELAs were 95.4, 94.2 and 94.8 per cent.

As the metropolitan areas have grown, they have decentralised. In other words, their growth has been disproportionately concentrated outside their central cores, and in some cases these cores have experienced absolute decline of population and even, in a few cases, of employment. In 1931, out of a total population of 30.76 million in the 100 SMLAs of England and Wales, some 21.97 million (55.0 per cent of the total population of all England and Wales) lived in their core areas and 8.79 million (22.0 per cent of the total) in their rings. By 1951 these proportions had already changed to 50.5 and 26.8; by 1961 they were 48.3 and 29.2; by 1966 they were 46.2 and 30.9. It is possible, even, to conceive a day when the majority of people in the metropolitan areas might live in their rings.

One of the more significant findings is that employment has tended to be more concentrated in the cores of the SMLAs than has population. In 1951 (no figures being available for 1931), 58.3 per cent of all employment in England and Wales was found in the SMLA core areas. This proportion actually remained constant in 1961 and fell marginally to 56.8 per cent in 1966. Meanwhile, the proportion of employment in the rings had risen, from 20.7 in 1951 to 21.1 in 1961 and 22.8 per cent in 1966. Thus it is possible to conclude quite definitely that in relative terms, employment was decentralising modestly between 1961 and 1966, though far less strikingly so than population. Between 1951 and 1961, indeed, many individual metropolitan areas showed a tendency for relative centralisation of employment.

The 100 metropolitan areas are very unevenly distributed over the face of England and Wales, as Figures 4.1 and 4.3 show. They fall into two great groups including over four-fifths of them, together with one smaller concentration and a handful of isolated metropolitan areas in the peripheral parts of the country. One of the two great groups, numbering twenty-six metropolitan areas, is concentrated in and around London, and stretches up to 96 kilometres (60 miles) in certain directions. It includes the biggest of all metropolitan areas – London – and a whole host of rather small metropolitan areas based on the towns around the capital, which have developed as employment and commuting centres in their own right, so limiting the spread of London's commuting field, especially to the north and west. Because so many of these areas are small, the whole group accounts for a smaller proportion of the metropolitan population – 20 per cent – than of the total number of metropolitan areas. This major group has outliers, nearly, but not quite contiguous with it; one is a loose grouping of three metropolitan areas in South Hampshire, another is an equally loose group stretching up the East Anglian coast towards Great Yarmouth.

A second and even bigger group of metropolitan areas, numbering 40 out of the 100, extends across the industrial heart of England, including the East and West Midlands, Lancashire and Yorkshire. Even on the more restrictive SMLA definition, it covers all the main industrial axis of the Midlands and North, save for a small section of the South Pennines; the wider MELA definition covers the area completely (Figure 4.2), and joins it up with the first to give a continuous cover from Lancashire and Yorkshire down to the South Coast. Because this group contains so many of the larger metropolitan areas based on the great provincial cities – Birmingham, Liverpool, Manchester, Leeds, Bradford, Sheffield, Nottingham, Leicester and others – it accounts for a bigger share of population than of individual metropolitan areas: 43 per cent in 1961. It throws out tongues which extend

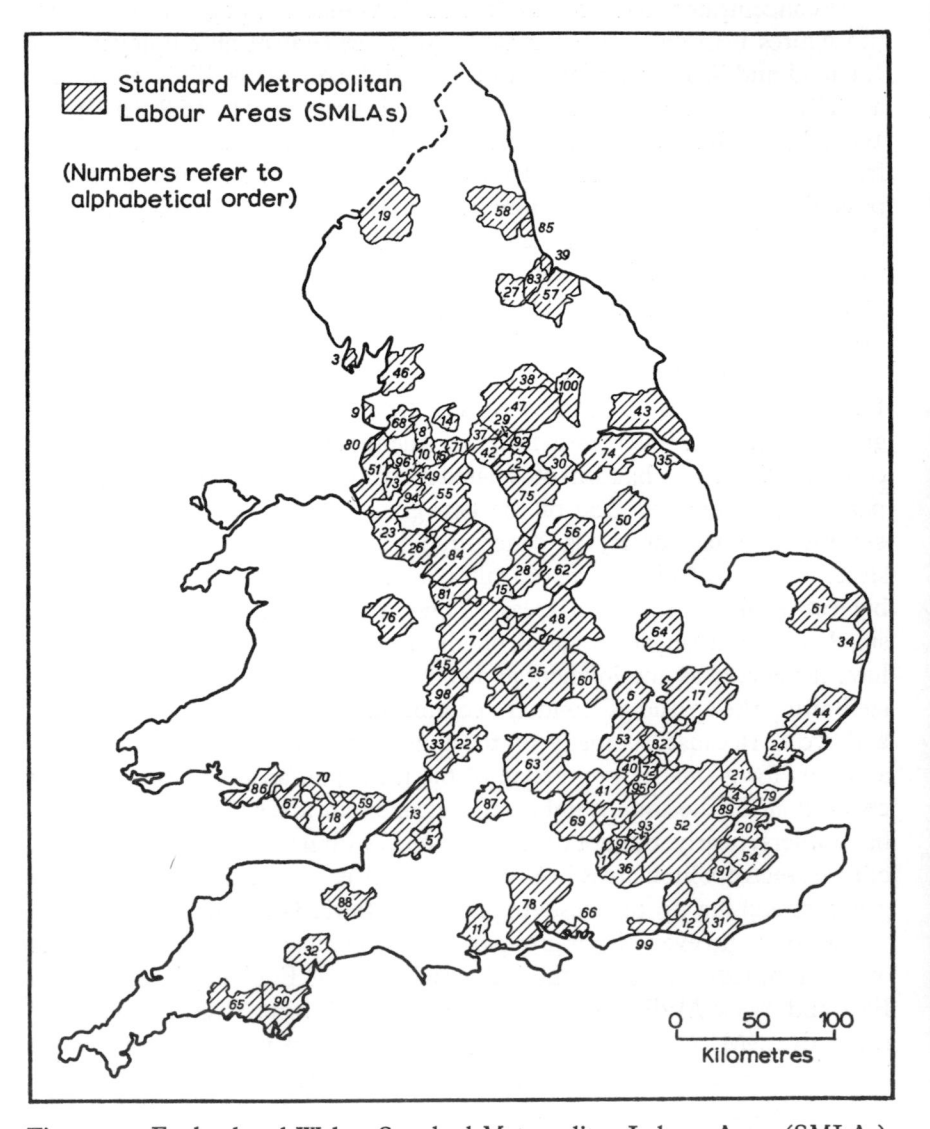

Figure 4.1. England and Wales: Standard Metropolitan Labour Areas (SMLAs), 1961. Key on p. 111.

Key to Figs. 4.1 and 4.2

Standard Metropolitan Labour Areas and Metropolitan Economic Labour Areas, England and Wales 1961

1	Aldershot	51	Liverpool
2	Barnsley	52	London
3	Barrow-in-Furness	53	Luton
4	Basildon	54	Maidstone
5	Bath	55	Manchester
6	Bedford	56	Mansfield
7	Birmingham	57	Middlesbrough
8	Blackburn	58	Newcastle-upon-Tyne
9	Blackpool	59	Newport
10	Bolton	60	Northampton
11	Bournemouth	61	Norwich
12	Brighton	62	Nottingham
13	Bristol	63	Oxford
14	Burnley	64	Peterborough
15	Burton-on-Trent	65	Plymouth
16	Bury	66	Portsmouth
17	Cambridge	67	Port Talbot
18	Cardiff	68	Preston
19	Carlisle	69	Reading
20	Chatham	70	Rhondda
21	Chelmsford	71	Rochdale
22	Cheltenham	72	St Albans
23	Chester	73	St Helens
24	Colchester	74	Scunthorpe
25	Coventry	75	Sheffield
26	Crewe	76	Shrewsbury
27	Darlington	77	Slough
28	Derby	78	Southampton
29	Dewsbury	79	Southend
30	Doncaster	80	Southport
31	Eastbourne	81	Stafford
32	Exeter	82	Stevenage
33	Gloucester	83	Stockton-on-Tees
34	Great Yarmouth	84	Stoke-on-Trent
35	Grimsby	85	Sunderland
36	Guildford	86	Swansea
37	Halifax	87	Swindon
38	Harrogate	88	Taunton
39	West Hartlepool	89	Thurrock
40	Hemel Hempstead	90	Torquay
41	High Wycombe	91	Tunbridge Wells
42	Huddersfield	92	Wakefield
43	Hull	93	Walton and Weybridge
44	Ipswich	94	Warrington
45	Kidderminster	95	Watford
46	Lancaster	96	Wigan
47	Leeds	97	Woking
48	Leicester	98	Worcester
49	Leigh	99	Worthing
50	Lincoln	100	York

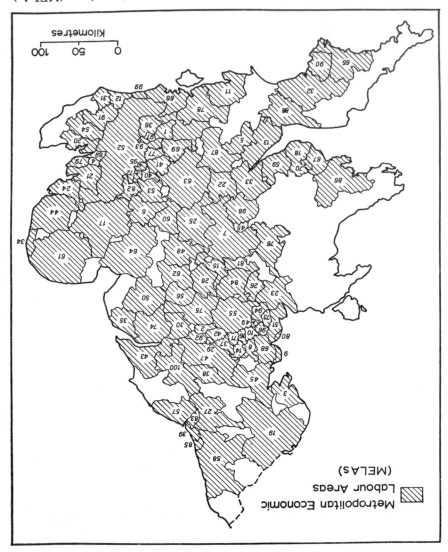

Figure 4.2. England and Wales: Metropolitan Economic Labour Areas (MELAs), 1961. Key on p. 111.

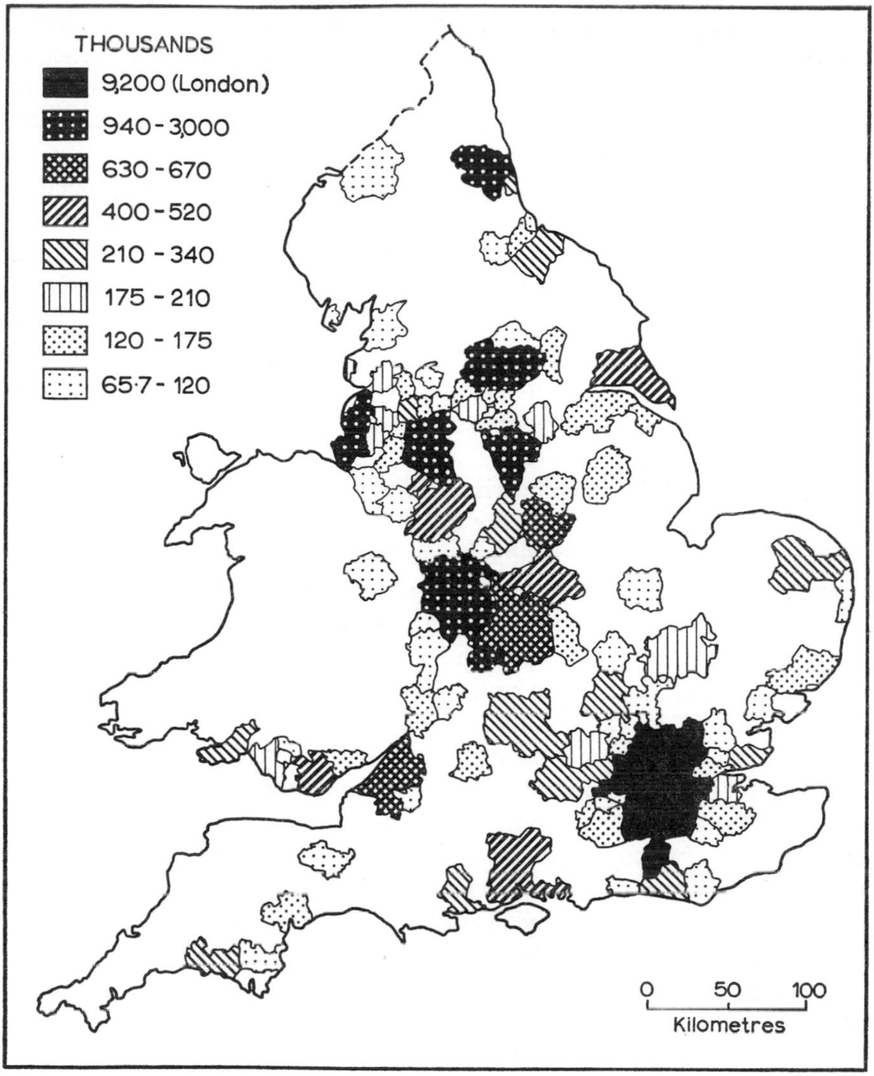

Figure 4.3. England and Wales: population of Standard Metropolitan Labour Areas, 1961

to Hull in the north-east and Gloucester in the south-west. And almost contiguous with it, and in fact forming extensions of it, are isolated metropolitan areas like Blackpool, Lancaster, Burnley and Lincoln, together with five SMLAs forming a distinct group in industrial South Wales, and Bristol plus Bath on the opposite side of the Severn estuary.

These two groups and their extensions and outliers, therefore, together account for over 80 per cent of the metropolitan areas of England and Wales, and for over 90 per cent of the population within them. Only one other area of concentrated metropolitan development exists; it is the North East industrial belt in Northumberland, County Durham and the North Riding, containing six further metropolitan areas. The remaining metropolitan areas of England and Wales, as Figures 4.1 and 4.3 show, are scattered widely round the peripheral parts of the country; they tend to be small in population.

The remainder of the comparisons in this chapter will be made, for the sake of simplicity, in terms of the more restrictive SMLA definition.

The one hundred metropolitan areas have had very different records of growth during the nineteen-fifties and sixties, whether this is measured in absolute or in percentage terms (Figures 4.4 and 4.5). About two-thirds of the total growth in all metropolitan areas in the nineteen-fifties (1,185,000 out of 1,889,000) and rather under one-half in the early nineteen-sixties (292,000 out of 640,000) occurred within the South East and the Midlands. In the South East, though, the growth transferred itself progressively from London to the surrounding ring of metropolitan areas. While in the nineteen-fifties London gained 44,300 people and its peripheral metropolitan areas gained 573,100, in 1961–6 London lost 266,000 and the peripheral areas gained 302,000. These peripheral SMLAs were among those displaying the highest percentage rates of growth, in both periods: 17 out of 20 fastest-growing SMLAs during the period 1951–61, and 15 out of 20 during 1961–6, were in the south East. A rather similar process occurred round Manchester, but here the peripheral SMLAs failed to take up the loss of population from the central SMLA, because of the generally weak population growth of the entire North West region. In general, the metropolitan areas with the poorest growth areas, both in the nineteen-fifties and sixties, were in the North West and in Yorkshire.

The regional effect, then, was one very strong explanation for the differences in population growth (or decline) among different metropolitan areas. But the other main factor was undoubtedly size. The biggest metropolitan areas all showed relatively poor growth rates, standing in the second half of the league table of growth. (One exception was Birmingham in the

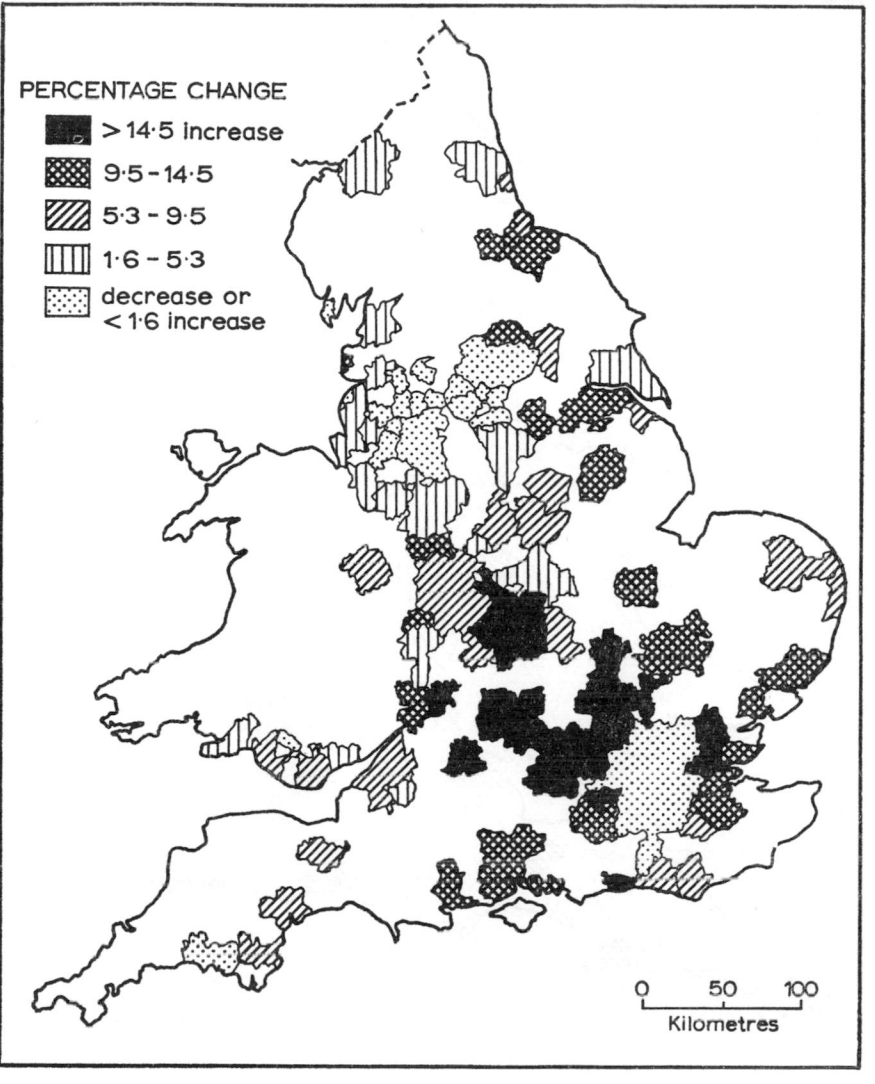

Figure 4.4. England and Wales: percentage population change by Standard Metropolitan Labour Areas, 1951–61

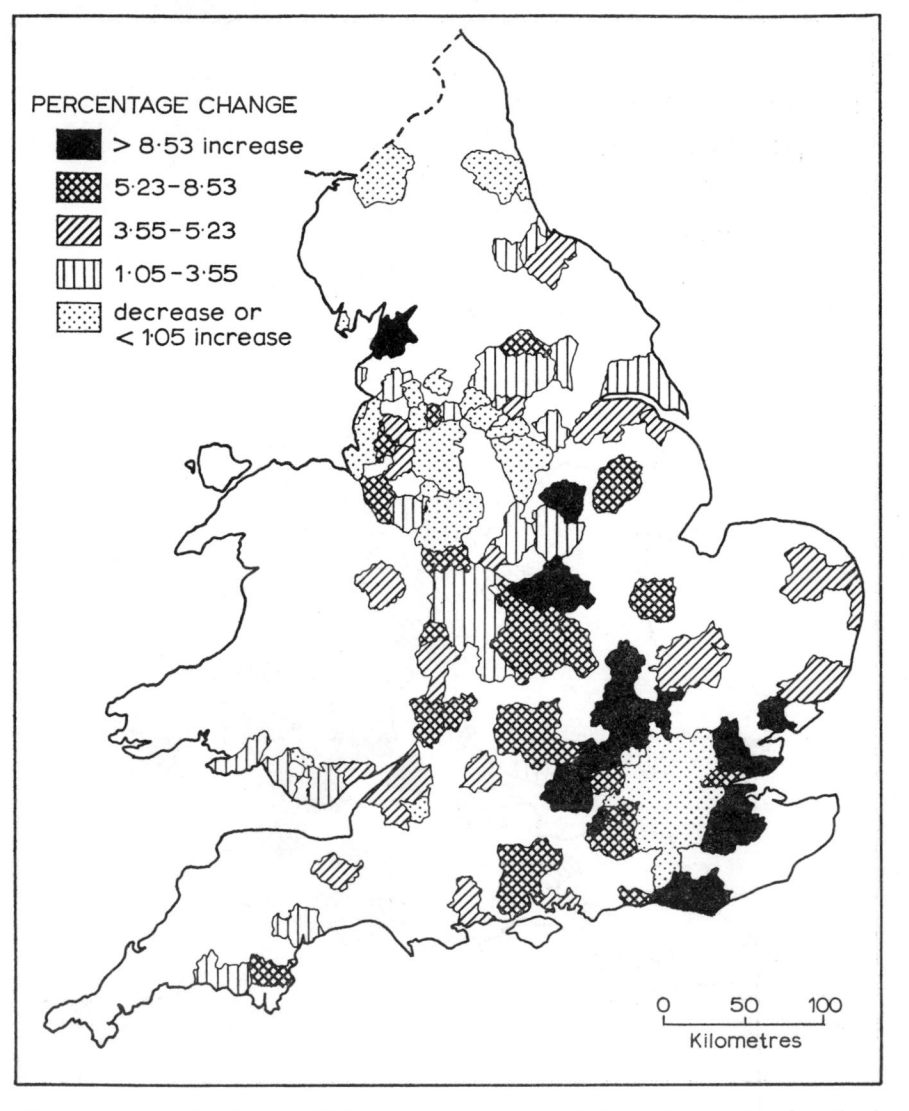

Figure 4.5. England and Wales: percentage population change by Standard Metropolitan Labour Areas, 1961–6

nineteen-fifties; it stood 48th.) In the nineteen-sixties, four major SMLAs showed actual declines: London, Liverpool, Manchester and Newcastle. Contrary to the popular legend of its dynamism, the London metropolitan area – biggest by far of the hundred – was 85th in the league table in the nineteen-fifties, and an ignominious 98th during 1961–6.

In general, employment tended to grow faster than population in most SMLAs, reflecting the fact that a progressively greater proportion of the potential adult labour force – especially the women – were being drawn into active work during the nineteen-fifties and sixties. The relationship between population growth and employment growth is less clear-cut for the nineteen-sixties than for the nineteen-fifties, and there are some individual discrepancies which are not readily explained.

Much of the interest of the analysis lies in the internal shifts of population and employment within each metropolitan area. Here a simple typology is necessary. Restricting ourselves as before for the sake of simplicity to the narrower SMLA definition, and comparing the cores with the rings, we can first separate out a minority of declining areas, and then divide the growing areas into the following groups:

Absolutely centralising: core increasing, ring decreasing.
Relatively centralising: both core and ring increasing, core faster than ring.
Relatively decentralising: both core and ring increasing, ring faster than core.
Absolutely decentralising: core decreasing, ring increasing.

Taking population first, we find that during the period 1951–61 no less than 87 out of 100 metropolitan areas were growing (Figure 4.6). The exceptions, of course, were mainly in the north. Of the 87, only 28 were centralising (5 absolutely); 59 were decentralising (19 absolutely). The centralising SMLAs were clearly a special case; they were disproportion-ately concentrated in the South East, and could be equated with the very rapidly-growing ring of small SMLAs round London. Here, both in the metropolitan areas based on new towns and in those based on private develop-ment, growth was very rapid and much of it took place in the central town of the SMLA. Absolute decentralisation, in contrast, tended to be associated with the bigger metropolitan areas of the north, where slum clearance was already beginning to reduce the populations of some central core cities.

The nineteen-sixties showed a sharp contrast (Figure 4.7). Again 87 metropolitan areas showed population growth, the exceptions being either declining industrial areas (generally in the north) or very large metropolitan

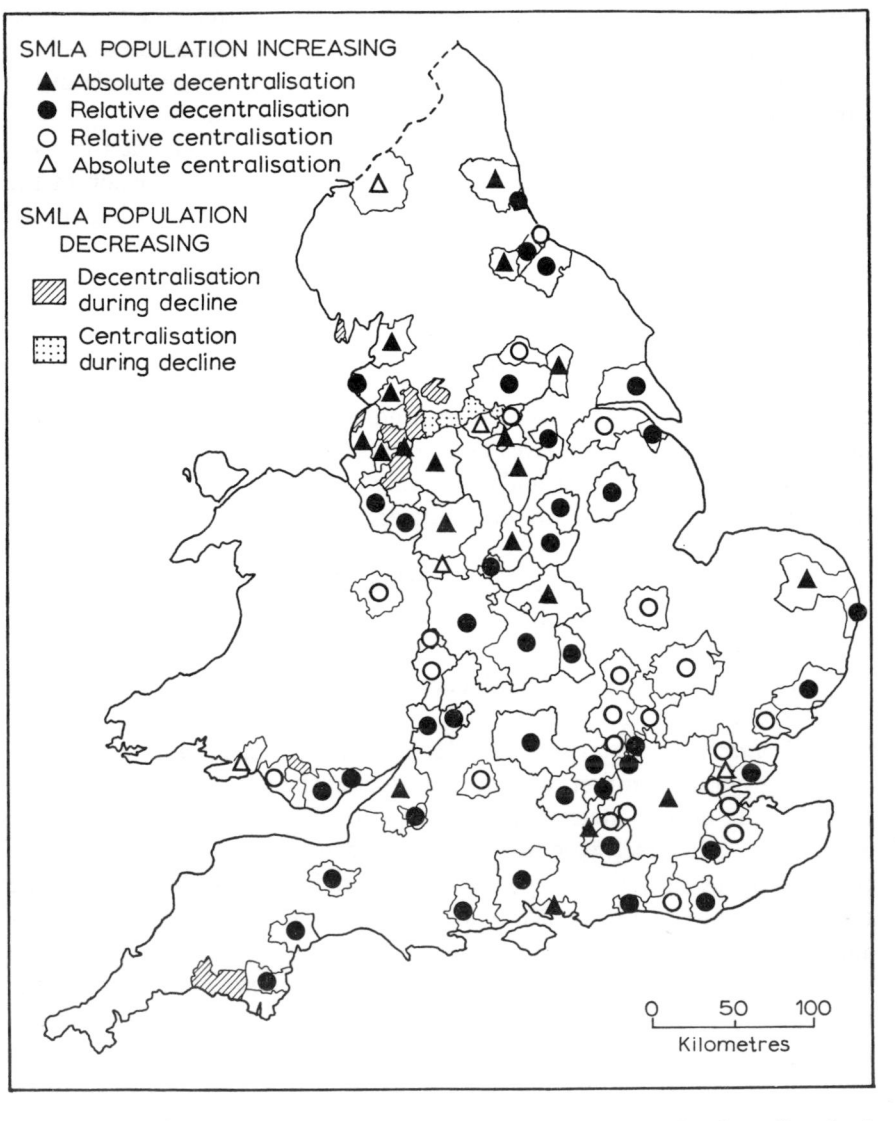

Figure 4.6. England and Wales: trends in population decentralisation from Standard Metropolitan Labour Area cores, 1951–61

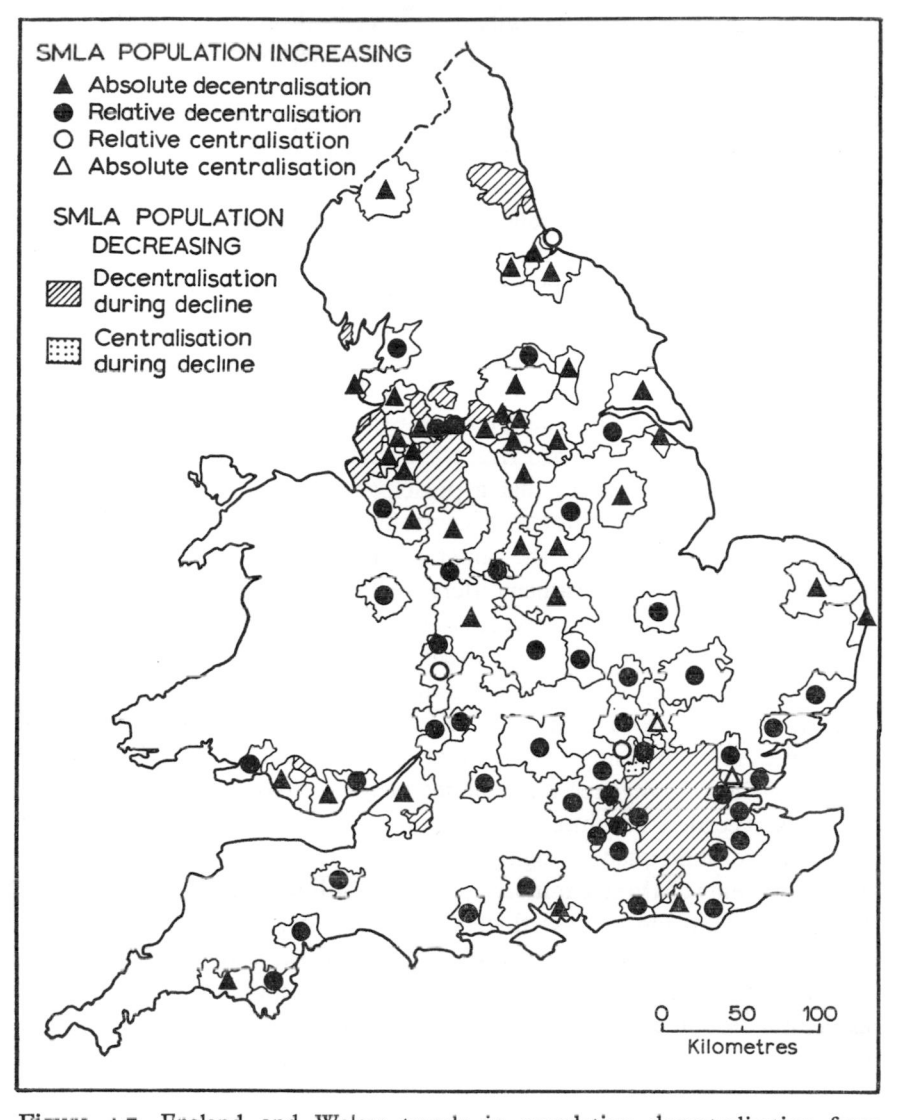

Figure 4.7. England and Wales: trends in population decentralisation from Standard Metropolitan Labour Area cores, 1961–6

areas, or both. But of the 87, only 5 were centralising (all relatively); 82 were decentralising, and 36 of these were decentralising population absolutely. Absolute decentralisation of population was again concentrated in the north, where slum clearance was reducing the population of the central city and leading to overspill into the ring. In Birmingham, Leeds, Nottingham and Cardiff, relative decentralisation in the nineteen-fifties turned into absolute decentralisation in the early nineteen-sixties. But, as already noted, some of the bigger metropolitan areas were a stage further advanced in this process; there, absolute decentralisation in the nineteen-fifties turned into decline of the whole SMLA during 1961–6. The relatively decentralising SMLAs, which constituted the great majority during the 1960s, in contrast tended to be the smaller and medium-sized ones, where there was still room left for expansion in the central core but where growth in the suburban ring was faster.

There is no doubt, then, that a principal result of the analysis is the strong and growing tendency to decentralisation of population in English metropolitan areas during the nineteen-fifties and early sixties. With employment, it has been rather different. During the period 1951–61, 81 metropolitan areas showed increases in employment (the exceptions, again, being mainly in the north); 46 of these showed centralisation (25 of them absolutely), while 35 showed decentralisation (only 2 absolutely). At this time, then, while a majority of SMLAs were decentralising people, a majority were centralising employment. Most significantly, no less than a quarter of SMLAs registered actual declines in employment in their rings, and thus recorded absolute centralisation. But the early nineteen-sixties showed a profound change. Eighty-six metropolitan areas showed increases of employment; of these a minority, 32, were still registering centralisation (2 absolutely), while no less than 54 now registered decentralisation (2 absolutely). We can safely generalise, then, and say that the tendency to decentralise employment occurred later than the tendency to decentralise residential population; it became a majority phenomenon only in the period after 1961. Looking at individual records, it is clear that there was a strong tendency for the larger metropolitan areas to decentralise employ-ment earlier and faster, though even there the pattern became stronger during the early nineteen-sixties. Even in the nineteen-fifties these large areas were exhibiting the trend, with one or two exceptions; after 1961 it became almost universal. But among smaller metropolitan areas, there are many examples where centralisation in the nineteen-fifties was replaced by decentralisation in the nineteen-sixties.

Looking at population shifts and employment shifts together, we can begin to develop a general model. Population tended to be decentralising from core to ring, both in the nineteen-fifties and sixties; a small but distinct group of centralising areas, associated with the rapidly-growing ring of peripheral areas round London, was observable in the nineteen-fifties but then disappeared. Among the bigger metropolitan areas, there seemed to be a sequence from relative decentralisation to absolute decentralisation to decline. Employment tended to centralise rather than decentralise in the nineteen-fifties, but then to decentralise after 1961. There was a particularly clear association here with size: the bigger SMLAs showed a strong tendency to decentralise jobs as well as residences even in the nineteen-fifties, and the tendency became stronger and more general after 1961. But other explanations for the rate of outward shift of jobs – growth rates, or concentration of employment in the core – do not prove very satisfactory when analysed. Size, it seems, is the key factor.

Lastly, it is possible to make a systematic attempt to compare trends in population and in employment. Because in many individual cases the trends are working in opposite directions for people and for jobs – centralisation of employment and decentralisation of population, for instance – it is not easy to make a statistical analysis; it is more satisfactory to put the information in the form of a matrix. Thus in the nineteen-fifties, the commonest cases were presented by relative decentralisation of people associated with absolute centralisation, relative centralisation or relative decentralisation of jobs – 39 cases out of a possible 100 combinations. In the early nineteen-sixties, the commonest cases were relative decentralisation of people accompanied by relative decentralisation of jobs, or absolute decentralisation of people associated with relative decentralisation of jobs (38 cases in all) – a significant shift towards decentralisation of both people and jobs.

Against this background, it is possible in conclusion to develop a very simple model of urban development. Most urban areas in England and Wales were still in the earlier stages of evolution of this model in the nineteen-sixties, but a few of the bigger metropolitan areas had reached the later stages.

In the first stage of this model, sometime in the nineteenth century, both people and jobs concentrate in the cores of the metropolitan area; the ring is strictly non-existent, but the area later to be incorporated in the ring is rural, in both a functional and a physical sense, and it loses population to the town.

In the second stage, roughly from 1900 to 1950 in many areas, and beyond that in the smaller areas, population begins to migrate from the core to the suburban periphery, because of greater mobility and the spread of owner–occupiership. But the population of the core continues to rise, so relative decentralisation is the general rule. Only in the bigger cities, before 1939, might slum clearance contribute to absolute decentralisation of people. Centralisation of employment, generally of a relative kind, is the general rule; for increasing office and retail employment at the centre is more rapid than new factory building at the periphery.

In the third stage, typical of larger metropolitan areas since 1951 but above all since 1961, there is marked absolute decentralisation of population, due to inner city redevelopment for commercial and slum clearance purposes. Though there are continuing increases in office and other service employment at the core, these are outweighed by the effects of redevelopment, and above all by the rapid growth of local service employment (plus perhaps some decentralised factory employment) in the suburbs. The result is relative – or very occasionally absolute – decentralisation of employment.

The fourth stage has so far been reached only by London, and perhaps by Manchester, during the nineteen-sixties. Necessarily, it applies only to very large metropolitan areas which stand at the centre of a system of metropolitan areas. In this stage, decentralisation both of people and jobs continues a stage further: the metropolitan area as a whole tends to lose both people and employment to other, peripheral, metropolitan areas or to other parts of the country (which of these two last depending on the general regional dynamism of the area). London and its peripheral ring of metropolitan areas formed the clearest example in the nineteen-fifties and sixties. One might well expect that in time, other large metropolitan areas – Birmingham, Manchester, Liverpool, Leeds – would display similar patterns. But it must constantly be borne in mind that each of these very large metropolitan areas has very distinctive characteristics – characteristics that do not emerge clearly from the very generalised analysis made here. To understand these differences, a different scale of analysis would be necessary. Metropolitan area analysis has its value, but it also has its limitations.

Some policy conclusions

The full-scale study by the team at Political and Economic Planning contains many additional findings based on the close study of local areas of rapid suburban growth, selected from a representative selection of SMLAs. It reaches further conclusions based on a systematic analysis of certain policies pursued by government since the end of World War II – particularly in the field of housing. The conclusions from these separate studies broadly reinforce those which emerge from this overview of geographical trends in metropolitan England. They may be summarised as follows.

First, there is an unmistakable tendency to suburbanisation, first of population and homes, followed at an interval by employment. In this respect Britain is following the American pattern, albeit with a delay of some years. Because population growth cannot be accommodated within the central cities at densities which will be politically regarded as tolerable, the tendency must continue. Its precise strength will however depend on the rate of population growth. If growth rates declined to levels approximating those of the nineteen-thirties, as seems feasible in 1970, the amount of suburban growth would be limited but it would not be stemmed.

The result will be that an increasing proportion of the population will lead a large part of their lives in suburbs. This is already true of a major part of the female and child population. As employment moves out, too, it will begin to be true of breadwinners. Factories, offices, shops, homes will be linked within a suburban system of activities. The central city will be used only for the occasional visit, perhaps to obtain a range of goods and services which cannot economically be supplied in suburbs. Because of the physical characteristics of the new suburban communities, and because of the relative affluence of the households living in them, movements will largely be by car, and public transport will experience considerable difficulties in running viably without subsidy. Car ownership, at least at the level of one car per family, should extend to 80–90 per cent of households by the nineteen-eighties. The evidence of the 1966 census shows that already the great majority of work trips originating in suburban areas are made by private transport, many by car.

Second, the form of this change will vary somewhat according to the region of the country and especially according to the size and form of the urban area. Very large metropolitan areas are likely to follow the London model, decentralising people and then jobs to independent labour market systems based on medium-sized towns, up to 64 or 80 kilometres (40 or 50 miles) from the centre of the parent agglomeration. Within the smaller metropolitan areas this will be less marked; employment will be scattered in the suburban ring, but a higher proportion may remain in the central city, with a stronger tendency to radial commuting, possibly by public rather than private transport.

Third, the suburbanisation trend may merely be the outward manifestation of a deeper-seated motivation: for sociological studies show that a large proportion of those settling in representative new suburbs, though recognising that they are now suburbanites, aspire and expect to live in the countryside eventually. At present, life in the countryside can be regarded as a peculiarly prestigious form of life for the ex-urban commuter or retired person; studies indicate that many other people want to join this small privileged group, and as they acquire wealth and political power their demands may become insistent. In the past, the PEP study shows that rural England has successfully resisted the demands of the great majority of urbanites, partly due to the historical accident that the system of planning authorities set up in 1947 recognised the historic division of town and country. With the reform of local government, it is difficult to see how long this defence could successfully endure.

The most important policy conclusion from the PEP study, then, is for the planning of rural areas in the future. In the past a largely negative policy of preservation against alien urban interests has been successful. For the future, a more positive policy will be needed. It will have to resolve the very real local conflicts between agriculture, afforestation, recreation, urban development, defence and many other uses. For this purpose, it will need more sophisticated analysis of the value of land in alternative uses – including estimates of the benefits obtained from that vague and over-worked word, amenity. It will also need to relate this approach to more imaginative positive design planning which seeks to accommodate urban growth in the countryside in a variety of settlement forms, while carefully preserving the land which is most valuable in other uses – whether agriculture, recreation or water supply.

Eventually, the impact of these trends on existing urban areas will also be profound. First in the bigger metropolitan areas, then progressively in smaller ones, employment will decentralise relatively and later absolutely out of the towns (and especially out of the congested town centres) and into the new suburbs, following the movement set earlier by the residential population. Probably the absolute loss in employment and in service levels will not be too great, at least where the metropolitan area as a whole is experiencing growth. But in order that falling rateable values at the centre can be matched by rising values at the periphery, the reform of local government on city regional lines will become an absolute necessity. Otherwise, the fate of English cities will be that of too many American ones – financial strain followed rapidly by economic and physical deterioration.

References

Berry, B. J. L. (1967). *Functional Economic Areas and Consolidated Urban Regions of the United States: Final Report of the S.S.R.C. Study*, Chicago (mimeo).

Bureau of the Budget (1964). *Standard Metropolitan Statistical Areas*, Government Printing Office, Washington, DC.

Carruthers, W. I. (1962). 'Service Centres in Greater London', *Town Planning Review*, XXXIII, 5–31.

Carruthers, W. I. (1967). 'Service Centres in England and Wales, 1961', *Regional Studies*, I, 65–81.

Christaller, W. (1933; 1966). *Central Places in Southern Germany (Die Zentralen Orte in Süddeutschland)*, Prentice-Hall (translated by C. W. Baskin).

Clark, C. (1951). 'Urban Population Densities,' *Journal of the Royal Statistical Society*, A, CXIV, 490–6.

Clark, C. (1967). *Population Growth and Land Use*, Macmillan.

Clawson, M. (1971). *Suburban Land Conversion in the United States: An Economic and Governmental Process*, Johns Hopkins.

Davis, K. L. (1959). *The World's Metropolitan Areas*, California University Press (for international Urban Research).

General Register Office (1956). *Census of England and Wales 1951: Report on Greater London and Five other Conurbations*, HMSO.

Green, F. H. W. (1950). 'Urban Hinterlands in England and Wales: An Analysis of Bus Services', *Geographical Journal*, CXVI, 64–88.

Hägerstrand, T. (1970). Tidsanvändning och omgivningsstruktur. Bilaga 4 to *Urbanisieringen i Sverige (Bilagedel 1 till Balanserad regional Utveckling)*, Stockholm: Statens offentliga utredningar (SOU) (Official publications) 1970:14.

Hall, P. (1966). *The World Cities*, Weidenfeld and Nicolson.

Hall, P. et al. (1972). *Megalopolis Denied: The Containment of Urban England, 1945–1970*, vol. I. Allen & Unwin.

Kent County Council (1963). *Quinquennial Review, Report on the Survey and Analysis, Part Four*. The Council (4 vols.).

Lawton, R. (1959). 'The Daily Journey to Work in England and Wales', *Town Planning Review*, XXIX, 241–59.

Lawton, R. (1963). 'The Journey to Work in England and Wales: Forty Years of Change', *Tijdschrift voor Economische en Sociale Geografie*, LIII, 61–9.

Liepmann, K. (1944). *The Journey to Work*, Routledge.

Riley, R. B. (1967). 'Urban Myths and the New Cities of the South-West', *Landscape*, XVII, 1, 21–3.

Royal Commission on Local Government in England (1968). *Local Government in South East England*, HMSO. (Research Studies, 1, by the Greater London Group, London School of Economics.)

Schnore, L. F. (1962). 'Metropolitan Development in the United Kingdom', *Economic Geography*, XXXVIII, 215–33.

Smailes, A. E. (1944). 'The Urban Hierarchy in England and Wales', *Geography*, XXIX, 41–51.

Whyte, J. S. (1970). 'The impact of Telecommunications on town planning', *Town and Country Planning School, Swansea, 1970, Proceedings*, Town Planning Institute.

Wilson, A. G. (1969). *Disaggregating Elementary Location Models: CES-WP-37*. Centre for Environmental Studies (mimeo).

5. Poverty and the urban system

RAYMOND PAHL

The main purpose of this chapter is to explore one aspect of the relationships between certain long-term trends in the industrial and occupational structure of the population and the ecological structure of British cities. There are certain indications that the distinction between the skilled manual workers and the semi-skilled and unskilled manual workers is gaining a new significance in its urban spatial context. However, it is important to be clear at the outset that firm empirical evidence for the main outline of the argument is still lacking; a secondary objective of the chapter therefore is to argue for more research to be focussed upon this issue.

Changes in the industrial and occupational structure

It is initially worth adumbrating some of the well-established trends of changes in the country's occupational structure. Overall, it is evident that the total labourforce has been growing very slowly – at about 0.3 to 0.4 per cent per annum in England and Wales from 1931 to 1961. This rate of change masks more substantial changes at different levels, with more rapid rates of decline at the bottom of the occupational hierarchy and more rapid rates of increase at the top. The population as a whole is getting more skilled and the number of male semi-skilled and unskilled workers has declined absolutely in recent years (Knight, 1967). This decline in the proportion and number of less skilled workers is not taking place at the same pace between industries and occupational categories. Certain industries are showing a very rapid decline in the number of semi-skilled and unskilled male workers they employ. Thus, taking the railways as an example, from 1961 to 1966 the number of porters, ticket collectors and lengthmen in Great Britain declined by a third and the proportion of railway guards by just over a fifth. These losses were offset during the same period by increases in other sectors; e.g. office and window cleaners of 19 per cent, caretakers and office keepers by 11 per cent and hospital and ward orderlies by 7 per cent. In absolute terms, the railway occupations mentioned above declined by 35,000 and this was offset, for example, by

TABLE 5.1. *England and Wales: industrial structure of employment in 1951, 1961 and 1966, and Greater London in 1966*

Industry group	England and Wales			Greater London 1966
	1951	1961	1966	
	(percentages)			
Primary (Orders I–II)	8.6	6.4	5.2	0.2
Manufacturing (Orders III–XVI)	35.8	36.8	35.4	29.9
Construction (Order XVII)	6.3	6.8	7.7	6.9
Services (Orders XVIII–XXIV)	49.3	50.0	51.7	63.0
	100.0	100.0	100.0	100.0
Total (thousands)	19,940	20,913	22,325	4,430

Source: Greater London Council (1969), *Greater London Development Plan, Report of Studies*, Table 3.6.

the increase in the number of those employed as drivers of road goods vehicles alone (DEP, 1969*b*).

Such changes in the occupational structure reflect well established trends in industrial structure and there are indications that the rate of change is accelerating. Thus, from 1951 to 1961 workers in the manufacturing sector in England and Wales increased by 0.8 per cent per annum and those in the services sector increased by 0.6 per cent per annum. However, from 1961 to 1966, manufacturing workers increased by only 0.6 per cent per annum while numbers in the service sector grew by as much as 2.1 per cent per annum. Table 5.1 shows how the industrial structure became weighted towards the service sector from 1951 to 1966 and illustrates the overwhelming importance of the service sector within the Greater London Council area.

This growth in the service sector of the economy is accompanied by a dramatic increase in the number of less-skilled workers in service occupations. Table 5.2 shows that, for the four main service sector industrial orders in Great Britain between 1951 and 1961, the number of men and women in semi-skilled manual occupations grew by 15 per cent whilst the unskilled increased by 25 per cent. Workers with the least skills are being increasingly concentrated in the service sector of production. This trend is shown to an even more marked degree in a large urban area such as Greater London.

TABLE 5.2. *Great Britain: the growth of the proportion of less-skilled workers in the service sector, 1951–61 (males and females combined)*

	Occupation	
Industry	Semi-skilled manual	Labourers and unskilled
	(thousands)	
(1) xx. Distributive Trades		
1951	1,249.7	202.1
1961	1,394.7	271.1
% change	11.6	34.1
(2) xxi. Insurance, banking and finance		
1951	7.7	37.7
1961	6.7	49.8
% change	− 13.0	32.1
(3) xxii. Professional and scientific services		
1951	124.2	219.2
1961	146.9	383.4
% change	18.2	74.9
(4) xxiv. Public administration		
1951	140.6	331.0
1961	205.1	284.1
% change	45.9	− 14.2
(5) (1–4)		
1951	1,522.2	790.0
1961	1,753.4	988.2
% change	15.2	25.1
(1–4) as % of total occupational category: 1961	28.1	25.8

Source: Ministry of Labour (1967, Table 1A).

The low paid workers

Turning now to consider which are the lowest paid occupations, it is evident from Table 5.3 that the typical low paid worker is much more likely to be working in the service sector than in manufacturing or primary industry. Of the twenty lowest paid occupations in Great Britain, fifteen are in urban-based, service-sector jobs, probably in scattered work situations with a relatively underdeveloped Trade Union structure. Some indication that the lower paid workers are not increasing their wages as

TABLE 5.3. *Great Britain: the twenty lowest paid occupations,*
September 1968

		Full-time male workers earning less than £17 for a full week's work
		%
1	Gardener, grounds keeper	70.3
2	Caretaker, office keeper	66.2
3	Farm worker	61.9
4	Cleaner	54.1
5	Goods porter (not railways), materials mover (hand)	46.9
6	Shop salesman/sales assistant	46.4
7	Coalminer (surface)	45.9
8	Clerk – routine	45.3
9	Storekeeper, storeman, warehouseman or assistant – unskilled	41.1
10	Butcher, meat cutter	35.1
11	Guard/watchman	34.6
12	Male nurse, etc.	34.3
13	Labourer	34.0
14	Chef/cook	29.7
15	Storekeeper, storeman, warehouseman or assistant – semi-skilled	28.9
16	Packer, bottler, canner	28.4
17	Roundsman – retail sales	28.3
18	Postman, mail sorter, messenger	25.5
19	Storekeeper, storeman, warehouseman or assistant – skilled	21.4
20	Manager – retail shop	20.5

Source: 'Results of a New Survey of Earnings', *DEP Gazette*, May 1969, Table 4.

rapidly as more highly paid workers has been provided in reports prepared by Incomes Data Services Ltd (IDS):

In the last five years retail prices have risen by 24.7 per cent. During this period, in 25 out of 53 Wages Councils, the lowest minimum rate of pay for men has risen less than 24.7 per cent. Since September 1968, prices have increased by 12.1 per cent. In 30 of the 53 Wages Councils pay has gone up more slowly. Not only have the lowest paid become relatively worse off. [The differentials between the minimum pay rates of the lowest grade of workers in traditionally low paid industries and the lowest grade in more highly paid industries is significantly widening – IDS

Report 97.] In some cases, those actually on the minimum are worse off now than they were two years ago or five years ago. They have suffered an actual cut in their standard of living. Their real pay has decreased [IDS *Report 98*, 25].

It would be a task beyond the scope of this paper to explore in detail the social geography of the poorest paid workers, who are gradually getting relatively poorer (Sinfield and Twine, 1969). However, it would in general seem likely that such workers are in the declining industrial towns of the north and north-west and in the centres of the large conurbations. Certainly, it would not be a surprising conclusion to find that those with the lowest pay were living in the worst urban environments. This is most clearly confirmed in evidence provided by the London Housing Survey undertaken by the GLC in 1967.

Over three quarters (77 per cent) of households in the high stress area have a head of household in a manual occupation compared with just under one half (45 per cent) of those in the low stress areas. This is accentuated in both the 'service and semi-skilled' and 'unskilled' groups, where, although the total numbers in the groups are nearly equal for the two areas, they account for 34 per cent of total households in the areas of high stress, but only 10 per cent in the areas of low stress. With such a difference in socio-economic groupings it is hardly surprising that lower incomes prevail in the areas of high stress. One half (53 per cent) the heads of households in the low stress areas have a weekly income of £20 or more, but this applies to only 23 per cent of heads of households in the high stress areas [GLC, 1970, 52].

It must be made quite clear that direct connections between these data cannot be readily made. Thus, it is not possible to assert categorically that broad shifts in the industrial and occupational structures are leading to a growth of low paid workers in the service sector, concentrated in city centres. There are certainly some *indications* that this is taking place and there are stronger indications that the differential between the lowest-paid manual workers and better-paid manual workers is increasing. However, it must be remembered that skilled workers in some industries earn much less than semi-skilled and unskilled workers in other industries. Also some unskilled workers in the service sector gain substantial additions to their stated income in the form of tips, though this may be offset by uncertainty and instability of employment. These issues are discussed in Marquand (1967), Sinfield and Twine (1968) and Atkinson (1969).

The most up-to-date evidence on earnings is provided in the most recent DEP Study (1970*b*). Some attempt to relate low pay to housing conditions in London was made in the supporting research for the *Strategic Plan for the South East* and the Report concluded 'Generally housing takes a higher proportion of income in London than elsewhere in the country...but the proportion spent by those with low incomes, who generally have the worst housing, is higher still and annually increasing' (South East Joint Planning Team, 1970, 25).

The example of the City of London

It is paradoxical that when the dominant trend in all cities in advanced industrial societies such as Australia, France, Sweden, the USA and Britain is that of dispersion and diffusion, with a general fall in population within administratively designated urban areas, the poor may be concentrating and may be becoming poorer – certainly relatively, and probably absolutely. Part of the reason for this localization is given in a recent economic analysis of the City of London, which emphasises the professional and financial concentration without acknowledging explicitly the similar concentration of service workers.

The most noticeable feature of the City of the future will be 'specialisation within specialisation'. Not only will the type of activities carried out be almost entirely in 'rights to goods'; *within* each activity only those functions involving decision taking and close contacts with people – clients, customers or competitors – will remain.... The inter-related network of activities will be even more closely associated with each other [Economists Advisory Group, 1968, 28].

This same research group did a random sample survey of firms and gathered detailed information on all the communications made by 526 respondents in senior positions. In all, these people collectively took part in over 28,000 communications in one day. Of these communications 1,482 or nearly three per head were carried by special messengers (Dunning and Morgan, 1970). A quarter of all male postmen, mail sorters and messengers in 1968 in Great Britain received less than £17 for a full week's work and a half received less than £20 (DEP, 1970, Table 4). In the City of London in 1966 there were 3,610 postmen and mail sorters and 6,440 male messengers. These 10,000 workers are clearly just as concentrated as the 11,230 professional accountants, company secretaries and registrars who also worked in the City at that time (General Register Office, 1968, *Sample Census 1966: Economic Activity Tables*, 51–2). The continued development of the City as a financial centre is confidently expected (Economists Advisory Group, 1969) and if Britain enters the European Economic Community then, in *The Times*' words (23 October 1970), 'the City of London will become the Wall Street of western Europe'. Unfortunately there appears to be no research which can give the multiplier effect of increasing financial and other activities in the City. There are 70,000 clerks and cashiers in the City but only 22,000 administrators and managers; there are over 10,000 men and women serving and working in the catering trade as waiters, counter hands, cooks and kitchen hands; there are over 10,000 men and women working as shop assistants or managers; there are over 7,000 men and women working as warehousemen, storekeepers, packers and bottlers; there are 6,000 female office cleaners; and so on. Admittedly of these 33,000 service workers some 13,500 are married women working part-time but even if we limit our discussion to the 19,000 full-time male workers and add the postmen and messengers, the total is considerably greater than the total of administrators and managers in the City.

Quite evidently the City of London is an extreme example to choose: its characteristics are closer to those of other world cities, such as Paris or New York, than to those of other cities in Britain. However, it does highlight the point that increasing specialisation and increasing concentration of highly qualified professional and managerial workers carries with it a similar concentration of service workers, some of whom appear to be in the lowest paid categories. The City is also exceptional in that the concentration is extreme, with some 360,000 economically active people in 1966 on a mere 274 hectares (677 acres). Clearly, other London Boroughs house all of these people except the small minority of workers who commute from outside the Greater London area or who actually live in the City. If, as appears to be the case, the routine and non-specialist office functions can be more readily decentralised to suburban situations and beyond, then evidently the changing occupational structure of the City will have important implications for housing and the provision of other urban resources and facilities. Thus it is not surprising, for example, that no messengers live in the City, but that in 1966 over 5,000 lived in the three boroughs of Islington, Hackney and Southwark, which immediately adjoin the City and include some of the very worst housing conditions in the whole of London (Standing Working Party, 1970, Table 2).

A cautionary note must be added here: it is not possible to show that those low-paid workers who are becoming relatively poorer are necessarily the same people who work in the City and that, in turn, these people are the same as those who live in areas where the housing conditions present a problem 'more acute and intractable than any to be found in the remainder of the country' (Central Housing Advisory Committee, 1969, 141). Until further detailed research based on surveys of the incomes, occupations and workplaces of those in bad housing conditions is carried out, we are left with inferences based on fragmentary evidence.

The redistribution of real income and the equitable distribution of public services and facilities

It has been stated that there may be a process of pauperisation engendered by changes in the country's industrial and occupational structure. It is necessary to consider, therefore, other sources of real income apart from the labour market.

The orthodox view that life chances are directly related to income gained from the work situation alone is perhaps too limited. Thus, for example, wage differentials might have been decreasing but the poor could still be getting relatively poorer if their greater income were relatively less valuable as a means for getting scarce urban resources and facilities. Differentials in *real* income could increase, as differentials in wages decrease. Hence the operation of 'the urban system' could lead to a process of pauperisation independently of the situation in the job market. In Eastern Europe, attempts to redistribute resources in favour of less-skilled workers were frustrated by the distributional effects of an urban system which appeared to be creating new or deeper patterns of inequality (Konrád and Szelényi, 1969). It is evident that even in a socialist economy the inability to understand urban distributive mechanisms can inhibit the move to greater equality, based on earned income alone. It is evident also that the availability of some of the more important urban resources and facilities is not dependent solely on the ability to pay. Those provisions which are administered by local or national government are critical determinants of life chances. Such facilities may be compulsory – such as primary or secondary education – or may be provided upon request as a right or may be granted as a privilege. One strand in the ideology of the Welfare State as a type of society was that need should be a more important criterion than wealth to determine access to facilities. If this ideology had been comprehensively pursued and supported, we might have expected a considerable redistribution of real income within and between localities. Thus, to take the example of the physical distribution of primary schools, the state attempts to provide schools in relation to the age structure of the population, the pattern of residential development and the distance a small child can be expected to walk. This contrasts with schools in the private sector which need not bear any relation to catchment areas but which rely on the wealth of their clients to transport pupils, often to inconveniently situated converted mansions.

If the provision of public services followed such a principle systematically, we would expect to find a positive correlation between the need for public services and facilities and their provision. Thus, the Welfare State and notions of citizenship would be a reality and the inequalities following from wage differentials would be compensated for, so that the poor would not be doubly penalised. However, it is a commonplace to observe that this does not happen, although detailed documentation presents many complicated methodological problems (Davies, 1968). The provision of public services and facilities has its own pattern of inequalities. It can be argued that the exploration of the systematic structuring of such inequalities provides a useful focus to students of stratification working in urban areas (Pahl, 1969). The pattern of territorial injustice exemplified in the work of Davies for Britain can be found in every industrial society (e.g. Scardigli, 1970) and the elimination of such inequalities could well be extremely difficult, even in a post-revolutionary situation. The hidden mechanisms of redistribution operate in socialist as well as in capitalist societies: one of the tasks of a radical sociology might be to expose such mechanisms and to consider how an understanding of the urban redistributive system can be used to develop new theories of stratification in industrial society.

Regard should not simply be paid to those services which are traditionally discussed by students of social administration and social policy – what might be called the Seebohm services (Seebohm Committee, 1968). It is also important to consider all other aspects of public policy which have a redistributive effect. The siting of power stations, hospitals or motor-ways all help to redistribute real income. Externality effects, as economists call them, are an important element in the system. And the central problem is to understand how policies and programmes designed to reduce in-equalities may actually increase them. It is a matter of exploring the distri-bution of 'fringe benefits generated by changes in the urban system', to take Harvey's phrase (Harvey, 1971).

The public provision of health and welfare services, roads, public transport, fire and police protection, libraries, schools, swimming pools, parks and so on, is financed out of public rates and taxes 'for the community at large'. No one openly denies that the poorest man has as much right to walk in the park, use the public library, call out a fire engine or have domestic help under certain circumstances, as the richest man. If this does not happen we must ask whether the public sector generates its own system of inequality, or whether it simply reflects the economic system within which it is encapsulated.

Unfortunately, a systematic sociology of public provision has hardly started, despite the contention of one urban economist that 'local public services bid fair to become the chief means of income *re*distribution in our economy' (Thompson, 1965, 118; present author's emphasis). Only educational provision has been systematically studied, although something of a start in the sociology of the Health Service and the physical planning process has been made. If we compare the amount of effort and energy that has been devoted to the sociology of industry (mainly private) with that devoted to the sociology of public services, the contrast is striking. Certainly something is known of *The Faith of the Counsellors* (Halmos, 1965), the ideologies of teachers, planners and social workers; but it is not very helpful to talk of 'bourgeois' values in these professions if very similar values are found in socialistic societies. Thus, for example, planners in both socialist and capitalist societies may share similar goals and values, even if the means at their disposal are different. The same applies to many other caretaking professions and to the managers of the public sector. To some extent, their power and ideology derive from the political party or council that employs them. However, it would surely be naive to assume that managers of the urban system are merely tools of their political masters, ready to shift policies and programmes according to political fashion. Clearly, there is a considerable tension between administrator, professional and politician which the Fulton (1968) and Maud (1967) Reports document for National and Local Government respectively. Such managerial tensions are likely to produce different patterns of resource allocation in different localities as the distribution of power shifts between the three contending groups. This point is made by Davies:

The power of a Committee over the Council may be the result of many factors. The authority may always have prided itself on the performance of one service – perhaps having been a pioneer in this field – and be generous in its allocation of funds to it. A chief officer may have built up a distinguished team and have a distinguished successor who may ensure the continuation of the department's status in the authority...Members may be particularly sensitive to the needs of some services because of local pressure groups' activity, because of the political ideology of dominating members or because the needs of the area, as perceived by almost anyone, suggest that an above-average development of the service in question is appropriate. Such factors may vary greatly in strength between authorities, but there can be no doubt that they have a major effect on the extent and manner of provision of services [1968, 114–15].

There appears to be no systematic relationship between the level of provision of public resources and facilities and the political party in power. Old people, for example, do not necessarily receive better care under a Labour than under a Conservative local authority. There exists a need for a typology of allocative structures based upon a number of dimensions. At present it is known that territorial inequality is substantial and that men with the same occupation, income and family characteristics have substantially different life chances in different localities. The systematic portrayal of such inequalities remains to be undertaken. The range of territorial inequality is probably greater in many instances than the range of inequality between certain positions in the occupational structure; moreover, the former inequalities are widening more rapidly than the latter. Authorities differ in the rules they adopt for the allocation of public services and facilities; they differ in wealth; they differ in the number and the degree of skills of the professional gatekeepers they employ; they differ in the system of administration and organisation which they adopt and the degree of cooperation and coordination which may exist between 'chief officers' and the rest of their managerial staff. Max Weber was surely wrong when he described the end of the political autonomy of the medieval city as the end of a distinctive urban sociology (Weber, 1960). There is still a considerable concentration of power over the allocation of scarce resources in localities. Although there may be, as it were, conflicts between the line and staff managers and the shareholders (i.e. the Council), there is a united and common ideology of the need to control, plan and provide for the lower participants or clients in the urban system. There is a vested interest in maintaining the Local Authority (a title with significant Dahrendorfian implications) (Dahrendorf, 1959) which pays the managers good salaries and provides the context for the expansion of various professions. To some extent, each professional group in an Authority's management structure is an interest group, which gains power by increasing the number of its own clients or lower participants. Since few Authorities have an overall programme or policy, such as may occur at the national level, the cumulative effect of *ad hoc* policies, reflecting a fluctuating power situation, may provide greater 'diswelfare' and increasing inequality for the majority of an Authority's lower participants.

Managers and professionals in the urban distributive system

Harvey suggests that 'it is tempting to hypothesize that any activity that generates strong external costs will be undercontrolled or undercompensated for' and that small, but well-mobilised pressure groups will be able to influence disproportionately allocational and locational decisions (Harvey, 1971). As an example he cites 'Central Business District Imperialism' in which the well-organised business interests of the central city (with their small-group oligopolistic structure) effectively dominate the looser and weaker coalitions found in the rest of the city. This view of the city is very clearly exemplified by the nature and style of objections to the Greater

London Development Plan, which formed the basis of an Inquiry which began in 1970. Individual London Boroughs engaged counsel to argue for more resources (floorspace allocations or population totals) and one of the longest and most elaborate cases for more resources was made by the institution which already had most – the City. Whatever the particular merits of the cases put forward, it is evident that the poorer areas and populations have less resources to compete for more resources. They are dependent on the advocacy of others.

Conflict, then, is a necessary condition of urban life:

In the most general terms, therefore, the city is composed of groupings of residents, clearly identified within a spatial location, who are competing for the resources that the city offers. Whether organised or not, they are competing for good jobs, good houses, good education, and all the other 'goodies' available in the urban milieu [Meltzer and Whitley, 1968, 173–4].

The same authors go on to argue that slums exist and are maintained because the people living there have insufficient influence or control over the allocation of resources.

The argument so far is that scarce urban resources and facilities are distributed by the managers of the urban system (elaborated in Pahl, 1969), and that those professionals who, directly or indirectly, manage these resources are concerned individually to advance as professionals and collectively to maintain the boundaries between themselves and other professional interest groups. The important question which now emerges is the degree of independent and autonomous control which can be exercised by the urban managers as distinct from industrial managers or national and local politicians. How far, in terms of the main theme of this chapter, can urban managers work within the existing economic and political system to make their limited urban system operate more equitably? Furthermore, what indicators would be appropriate to measure the effects of the redistribution of real income within urban systems?

This is a difficult issue and all that one can say with certainty is that it is easier to find examples where the differentials between the poorest and the richest social groups have increased, rather than the reverse. Some interesting material is presented in *The Property Boom* by Oliver Marriott, who was financial editor of *The Times* from 1945 to 1965. Discussing the development of provincial town centres between 1948 and 1955 Marriott notes:

Building licences were only granted for replacing bombed shops or for shops on new housing estates. Each local authority was given permission by the Ministry of Housing for a certain amount to be spent in its area, and if the allocation was not taken up by the year end, it lapsed. This spurred the local authorities to busy on with redevelopment and quickened the flow of work into Ravenseft's eager hands between 1950 and 1955. In these crucial years Freedman and Maynard had virtually no opposition. One reason was that they were opening up a type of redevelopment which had previously been unknown. This was large-scale cooperation between municipal authorities and private enterprise....

The grapevine of local officials also helped Freedman and Maynard; at their annual conferences the town clerks or the surveyors or the borough architects would discuss the developer, which for a few years meant a word of mouth advertisement for Ravenseft....

Towards the end of 1953 Freedman and Maynard saw that before long they would be running out of blitzed cities. They began to look at the New Towns... the multiple shops flocked in to the New Towns' centres – and so did the New Townspeople....

In 1953 Harold Samuel's land securities bought out Freedman and Maynard and friends and took its holding in Ravenseft from 50 per cent to 100 per cent. This deal valued Ravenseft at £2,100,000....

...by the summer of 1967, after the property share boom had cooled, Harold Samuel's holding in land securities was worth £12,400,000, Louis Freedman's £1,850,000 and Fred Maynard's £845,000. The entire company, including the much bigger stake of the public and the insurance companies, was valued by the stock market at £62,000,000 [Marriott, 1967, 61–5].

And that is how we got our nice new shops and Louis Freedman made a personal fortune of nearly £2 million. Certainly it was government policy to redevelop town centres with the profit and benefit of private developers in mind and local authorities could do little else than pass the best central sites to commercial organisations. Sites for public services and facilities had to be found away from the most accessible locations. There was little scope for policies for a more equitable distribution of real income.

The current interest in cost–benefit analysis as a managerial tool in urban resource allocation may also lead to a widening of real income differentials, since it has an inevitable bias towards the values of property and capital. In the Roskill Commission's study of amenity costs at a new international airport, it was found that the depreciation of property values near Heathrow and Gatwick was highly sensitive to the class of property. On average, for all noise levels, the percentage depreciation of high-class property was four times the depreciation of low-class property. Simply because their property values were so little affected, the poor were assumed to be little bothered by noise, and small or nil values were attached to their loss of amenity for the purposes of the cost–benefit quantification (Adams, 1970). Again, although there are techniques for determining the cost of, say, an urban motorway to the owners of property in its path, other costs, less easily quantified, are not taken into account. In a working-class area of rented housing, kin links and supporting patterns of social relationships may have created a form of social capital which, on one system of accounting, is more valuable to those who have it than the gardens of the middle class. The problem is how to quantify such a value. Since the middle class are more likely to be compensated and may also suffer less from forced mobility – they normally have the means to overcome physical distance and to maintain close social distance where this has been established – they are much less likely to lose from changes in the physical structure of cities. And not only would they suffer less if they were to be disturbed (because of capital compensation and so forth), but they are less likely to be disturbed, simply because their costs are more easily quantified and because they are more articulate and knowledgable about proposals that may affect them. Sections of the working class are much more likely to be at risk, partly because of their lack of property and partly because of their ecological position in the city. The relationship between ecology and equality is relatively unexplored.

An interesting study by Ornati, *The Transportation Needs of the Poor* (1969), demonstrates that the developing ecological structure of a city – in this case New York – has built into it certain constraints which lead to an increasing disjunction between poor peoples' homes and their workplaces. Public transport routes, originally developed to link middle-class residences to middle-class workplaces, greatly inconvenience the poor who move into old middle-class areas and have different transport needs. A consumer-oriented society is more concerned with transporting the poor to spend their money in central shopping centres, Ornati implies, than to opening up a wider range of employment opportunities. Even the cost of getting to shops may be considerable, and is well illustrated in this country by the case of Glasgow, where new flats on the urban periphery have few local shops and the people not only have to pay higher rents but are obliged to travel back to the area where they used to live, now redeveloped as a shopping centre. Such examples illustrate how differentials in real income may increase as a result, in this case, of the increased cost of physical access to urban resources such as employment or shops.

Spatial policy and social policy

A further theme which has been a focus of controversy in relation to Birmingham, Glasgow, London and many other cities concerns the decentralisation of manufacturing industry. The suburbanisation of manufacturing industry is a well established trend in the United States: in Britain the development of New Towns, specifically planned to ease the pressures on the main conurbations, has accentuated the trend. Those growth industries which are expanding the most rapidly tend to be those which have a high demand for skills, a high ratio of managers to managed and a favourable ratio of floorspace to worker. Space and skilled workers are the crucial locational factors and the peripheries of city regions appear to be ideal sites. Thus, in the case of the London Region, the most rapidly expanding area economically is in the Outer Metropolitan Region, immediately beyond the Green Belt (Pahl, 1965; South East Joint Planning Team, 1970). The effect of this peripheral expansion may be to limit the opportunities for social mobility by those in the central cities, hemmed in by middle-class suburbs and offered employment either in old-established, and possibly declining, manufacturing industries, where the wages are often low (e.g. the textile industry), or the service sector, where opportunities for training and advancement may be less good. This point links back to the occupational structure of the expanding functions of the City of London. In the past large cities were more likely to have higher rates of social mobility (Lipset and Bendix, 1959). Such a conclusion must now be severely questioned as urban diffusion 'decentralises opportunity'.

This possibility that cities, which are continuing to concentrate a particular type of activity at the centre, may cause blockages in social mobility for the offspring of sections of the population now working in the service industries, is a very serious issue. It is the main theme of 'Social Structure and Social Change', a chapter in *Studies Volume 2* of the South East Joint Planning Team's Report (1971 a). Certain of the points touched on in this chapter are also discussed in that chapter. Of particular interest are the positive policy proposals for a *Sector City* which are elaborated at length in Chapter 6 of *Studies Volume 4* (1971 b). Here some attempt is being made to understand the social and economic forces at work in the urban system and to manage and organise the system so that resources may be reallocated to those at present most deprived and also to create the conditions under which rates of social mobility may increase. The interplay between social, physical (land use) and economic planning is innovatory and much more detailed research and experience are needed. It seems clear, however, that management of housing and employment needs to be phased together and that arranging the amount of coordination required seems beyond the scope of the market. If it is hard to imagine the activities of a New Town Development Corporation being undertaken by private agencies, it is even harder to imagine the coordination of urban redevelopment, renewal and industrial re-location within a city being done for private profit alone.

In a society where so much of the urban fabric was created over a century ago, during a period of rapid industrial and urban growth, it is inevitable that some people will be living in conditions defined by the rest of society as intolerable. What does not have to be inevitable is that such populations also have limited access to job opportunities, schools, health services and other facilities, which are differentially distributed within the city. If those with low wages have to bear a disproportionate amount of the costs of noise, pollution and renewal of the urban fabric then they will be doubly penalised. As Harvey (1971, 273) aptly notes, 'almost all of this extensive literature has focussed on the allocation problems posed by externalities and very little attention has been paid to the distributional effects'. He goes on to remark that the 'best' distribution of incomes involves problems of equity not easy to resolve and hence frequently avoided. That there is frequently 'a quite substantial regressive redistribution of income in a rapidly changing urban system' (Harvey, 1971, 277) should be a matter for serious concern and one would expect this to be reflected in the political arena.

Spatial inequality and political consciousness

Until now lower participants in the urban system have not developed much consciousness of their common deprivations. There are, however, some signs of change:

For generations, the arena of action was more narrowly the workplace, the setting of production. Recently in the United States, low income persons have been organising to affect their rights to welfare and other forms of government services rather than to affect the economic market [Miller and Roby, 1970, 143].

Political consciousness may be slow to develop in the urban system because the mechanisms of redistribution are hidden. However, it may be that well-organised squatting and sporadic demonstrations against high rents, or intolerable noise from motorways, are the forerunners of more organised locality-based associations concerned with the basic issues of urban equity. Attempts to assess the relative costs and benefits of various locational decisions have had the unintended consequence of making explicit the fact that some people benefit and some lose. When it is manifest that the *same* category tends to benefit in a variety of circumstances and situations a more conscious pressure for an equitable distribution of urban resources and facilities may emerge.

Urban analysis may help us to reformulate our notions of the nature of our distributive system and to redefine lines of conflict based on constraints and limitations of life chances, which restrict the *use* of income and the access to rights and benefits rather than access to income alone in the labour market. It is possible that, in terms of access to urban facilities, the division between the skilled manual workers and the less-skilled manual workers may be becoming more significant than the division between the skilled manual workers and the routine non-manual workers. This might lead us to revise our notions of the conventional cutting points in the class structure.

The inequities and possible pauperisation generated by the operation of the urban system are extremely hard to document precisely. Although Harvey may well be right that the hidden mechanisms governing the redistribution of income 'seem to be moving us towards a state of greater inequality and greater injustice', such an assertion is extremely hard to substantiate. And even if the assertion were to be accepted it would be extraordinarily difficult to incorporate into a political programme policies that would ensure greater equity. It is hard to envisage a society which would accept such clear goals of equity and few would consider it legitimate to coerce, even without the use of armed force, to achieve a given pattern of distribution. We are left with arguments based on a sense of justice which assume that multiple deprivation is not part of a good society. The same people should not be at the bottom of all the hierarchies and receive the worst of all the public services and private markets. Yet even given a willingness to achieve such a measure of social justice, the means to that end are not easy to determine. Questions of costs and benefits involve value judgements as well as calculations, and conflicts, even between those who agree on the ends, are likely to increase.

From assertions through research to policies

It is evident that these matters are too important to leave to the assertions and hunches of academics, journalists and politicians. There are very important research problems with direct policy implications which urgently deserve attention. Perhaps mentioning some of these would strike an appropriately cautious note on which to conclude this chapter.

Firmer evidence is required on the relationship between the employment shift to the service sector in the economy as a whole and the relative growth of low paid occupations. More knowledge is required concerning the spatial distribution of low paid occupations in all sectors of the economy. This chapter has concentrated on the inner city and emphasised the need for more information on the multiplier effect of senior professional and managerial jobs, which are located in the city centre, on the growth in the number of the low paid. Further, we simply do not have adequate information to enable us to say with certainty whether those in certain low paid occupations are the same people who live in the worst housing and receive the least of urban resources and facilities. As an example, clear empirical evidence is not available on the nature and degree of the limitations on social mobility amongst less privileged populations in the inner city. It is not known, in any precise way, how far the activities of a professional group associated with the management of the urban system – town planners – have, in fact, helped to produce greater equity in the last twenty years. The example of the redevelopment of provincial shopping centres was chosen to indicate that differentials probably increased as a result of such activity. No doubt contrary examples could be found: but what would be the final balance? It is somewhat alarming that we do not know. Quite clearly, there is a spatial dimension to the problem of poverty which poses some of the toughest problems of social policy. It is probably impossible ever to achieve a truly equitable distribution of resources and facilities. However, it is important to distinguish between those problems that are associated with long-term trends in the economic and spatial structure of our society – such as the decline in agricultural labour or the financial concentration in city centres – and those that are due simply to the way resources are allocated in specific localities.

Perhaps the most hopeful development in recent years has been the move to a more rational system of local authority financial budgeting. This has forced some managers of the urban system to make explicit who is to benefit from the allocation of scarce resources and when. As the distributional mechanisms become more clearly understood, the possibility of having a more equitable distribution of urban resources becomes greater. This is partly because more information can be fed into the political process and partly because it is much more difficult to perpetrate inequities if this activity is clearly apparent to everyone. This belief that more information makes for greater equity puts a heavy responsibility on research workers. Whose problems do we tackle first?

References

Adams, J. (1970). 'Westminster: the fourth London airport?', *Area*, 2, 1–9.
Atkinson, A. B. (1969). *Poverty in Britain and the Reform of Social Security*, Cambridge University Press, Department of Applied Economics, Occasional Papers No. 18.
Central Housing Advisory Committee (1969). *Council Housing: Purposes, Procedures and Priorities*, HMSO, Ninth Report of the Housing Management Sub-Committee.

Dahrendorf, R. (1959). *Class and Class Conflict in Industrial Society*, Rout-ledge.

Davies, B. (1968). *Social Needs and Resources in Local Services*, Joseph.

Department of Employment and Productivity (1969*a*). *A National Minimum Wage, An Inquiry*, HMSO.

Department of Employment and Productivity (1969*b*). 'Employment changes in certain less-skilled occupations: 1961–1966'. *Employment and Productivity Gazette*, 308–11.

Department of Employment and Productivity (1970). *New Earnings Survey 1968*, HMSO.

Dunning, J. H. and E. V. Morgan (1970). Proof of Evidence to the Greater London Development Plan Inquiry on behalf of the Corporation of the City of London. Unpublished.

Economists Advisory Group (1968). *An Economic Study of the City of London, A Summary*, The Corporation of London.

Economists Advisory Group (1969). *The Role of International Companies*, Com-mittee on Invisible Exports.

Fulton Committee (1968). *Report on the Civil Service* (Cmnd. 3638), HMSO.

General Register Office (1968). *Sample Census 1966: Economic Activity Tables for Greater London*, HMSO.

Greater London Council (1969). Greater London Development Plan, *Report of Studies*, County Hall London.

Greater London Council (1970). *The Characteristics of London's Households*, County Hall London.

Halmos, P. (1965). *The Faith of the Counsellors*, Constable.

Harvey, D. (1971). 'Social processes, spatial form and the redistribution of real income in an urban system' in M. Chisholm, A. E. Frey and P. Haggett (eds.), *Regional Forecasting*, Butterworths.

Incomes Data Services Ltd (1970). *Incomes Data Report 97*, August, 140 Great Portland St, London W1.

Incomes Data Services Ltd (1970). *Incomes Data Report 98*, September, 140 Great Portland St, London W1.

Knight, R. (1967). 'Changes in the occupational structure of the working popula-tion', *Journal of the Royal Statistical Society*, cxxx, 3, 408–22.

Konrád, G. and I. Szelényi (1969). *Sociological Aspects of the Allocation of Housing: Experiences from a Socialist non-Market Economy*, Hungarian Academy of Sciences Sociological Research Group, Budapest.

Lipset, S. M. and R. Bendix (1959), *Social Mobility in Industrial Society*, Univer-sity of California Press.

Marquand, J. (1967). 'Which are the lower-paid workers?' *British Journal of Industrial Relations*, 359–74.

Marriott, O. (1967). *The Property Boom*, H. Hamilton.

Maud Committee (1967). *Report and Papers on the Management of Local Govern-ment*, HMSO.

Meltzer, J. and J. Whitley (1968). 'Planning for the urban slum', in T. D. Sharrard (ed.).

Miller, S. M. and P. Roby (1970). *The Future of Inequality*, Basic Books.

Ministry of Labour (1967). *Occupational Changes 1951–1961*. HMSO, Manpower Studies No. 6.

Ornati, Oscar A. (1969). *The Transportation Needs of the Poor*, Praeger.

Pahl, R. E. (1965). *Urbs in Rure. The Metropolitan Fringe in Hertfordshire*, London School of Economics, Geographical Paper No. 2, Weidenfeld and Nicolson.

Pahl, R. E. (1969). 'Urban social theory and research', *Environment and Planning*, 1, 143–53.

Pahl, R. E. (1970). 'The sociologist's role in regional planning', Chapter 14 in *Whose City?* Longman.

Scardigli, V. (1970). 'La fréquentation des équipment collectifs', *Consommation*, Annales du CREDOC, 1, 1–27.

Seebohm Committee (1968). *Report on Local Authority and Allied Personal Social Services* (Cmnd. 3703), HMSO.

Sharrard, T. D. (ed.) (1968). *Social Welfare and Urban Problems*, Columbia University Press.

Sinfield, A. and F. Twine (1968). *The Low Paid: The Employment Market and Social Policy*, unpublished ms. University of Essex.

Sinfield, A. and F. Twine (1969). 'The working poor', *Poverty*, Journal of the Child Poverty Action Group, 12/13, 4–7.

South East Joint Planning Team (1970). *Strategic Plan for the South East*, HMSO, for Ministry of Housing and Local Government.

South East Joint Planning Team (1971 a). 'Social structure and social change', in *Studies Volume 2*, HMSO for the Department of the Environment.

South East Joint Planning Team (1971 b). 'The social, employment and housing problems of London', in *Studies Volume 4*, HMSO for the Department of the Environment.

Standing Working Party on London (1970). *London's Housing Needs up to 1974*, HMSO, for Ministry of Housing and Local Government, Housing Report No. 3.

Thompson, W. (1965). *A Preface to Urban Economics*, Johns Hopkins.

Weber, M. (1960). *The City*, Heinemann.

6. Some economic and spatial characteristics of the British energy market

GERALD MANNERS

Over the last decade the primary fuel base of the British economy has been substantially widened. To the country's traditional and indigenous energy mainstay, coal, and its first modern partner, oil, has been added the power of the atom and of natural gas. In the process, and not withstanding the new-found hydrocarbon resources under the North Sea, the emotional reluctance of a time-honoured energy exporter for the first time to become dependent upon substantial energy imports has been to a large extent overcome. Meanwhile, the producer–supplier industries operating in the British energy market have steadily improved the efficiency of their production, processing, transport and distribution systems and have thereby come to enjoy falling real costs albeit in varying degrees. The market, as a consequence, has progressively developed a more competitive quality.

Setting the pace in these developments has been the oil industry. Taking advantage of the international industry's widening resource base – especially the development of low cost fields in North and West Africa – and the global persistence of excess production capacity relative to existing demands (Adelman, 1964 a, 1964 b), it has been able to procure its crude-oil supplies in a market which has been characterised by a downward pressure upon field prices. Despite the rise in the general price level, the 1970 price for Persian Gulf crude oil f.o.b. stood at about one-third of what it was twenty-five years earlier. In addition, the oil industry has relentlessly secured ever-greater crude-oil transport economies, especially through the use of larger oil tankers, constructed at steadily falling costs and available at progressively more advantageous charter rates. The c.i.f. price of crude oil landed at British refineries has thus exhibited a steady decline in current values and especially in real terms; between 1965 and 1969 the annual average value of imported crude and process oils entering the United Kingdom ranged between £5.97 and £7.55 per ton, whereas a decade earlier the comparable figures were £7.85 to £9.86 per ton (Ministry of

Technology, 1970). As national crude-petroleum imports rose ten-fold from 9.4 million tons to 92 million tons between 1950 and 1969, not only have new refineries been constructed (see Figure 6.3) but existing ones have been considerably expanded. Especially at Fawley (16 million tons annual distillation capacity in 1970), but also at the slightly smaller refineries of British Petroleum at the Isle of Grain (10 m.t.y.) and of Shell at Stanlow (11 m.t.y.) and Shellhaven (10 m.t.y.), substantial economies of scale have been exploited for the first time in British refinery operations (Pratten and Dean, 1965). Moreover, in distributing their refined products, the several companies serving the British market during the nineteen-sixties also exploited the economies of larger coastal vessels, rail cars and road tankers (Brunner, 1962; Hubbard, 1963). In addition, they reduced the number and increased the size of their depots and intermediate handling facilities, with a consequential saving in unit costs. By 1966, for example, the Esso Petroleum Company had only 68 terminals for oil distribution in the United Kingdom; eighteen years earlier, it had used 472 to handle only one-third of the volume of oil products.

During the same period, the gas industry has also been completely transformed. As a producer of secondary energy in the form of town gas, it vigorously developed from the late nineteen-fifties onwards a set of new manufacturing technologies in response to the increasing expense and the growing scarcity of coking coal, its main feedstock; in turn, non-coking coals, heavy fuel oil and light petroleum distillate came to be used as the primary fuel for gas manufacture (Manners, 1965), and the unit costs of production fell from about 6 pence* per therm – depending upon the size, age and loading of the conventional carbonisation plant – to about 2.5 pence per therm. (One therm equals approximately 25,000 kilo-calories.) Overlapping in time and in part superseding these developments was, first, the decision to pioneer the transport of liquified methane in refrigerated ocean tankers from Arzew in Algeria to a terminal at Canvey Island on lower Thamesside, with a delivered price of about 2.5 pence per therm; and, second, the discovery of North Sea natural gas (Odell, 1968). The interim agreement between British Petroleum and the Gas Council, signed in 1967, for the delivery of natural gas at a beach price of 2.1 pence per therm was followed in 1968–9 by the conclusion of large contracts with the international explorer-producer companies for approximately 100 million cubic metres (more than 3,000 million cubic feet) of 'sea gas' per day; under these twenty-five-year agreements, the beach price of gas fell to 0.85 pence per therm for gas taken above a load factor of 60 per cent. The

* In this chapter, all values are expressed in decimal currency.

unit costs of gas transmission and distribution also fell substantially during the nineteen-sixties following the development of improved pipe-laying techniques (which reduced the capital costs of new facilities) and the use of improved steels (which allow higher pipeline pressures and hence greater throughputs). And, of course, the gradual conversion of the distribution system to natural gas, with its higher calorific value, is in the process of doubling the system's capacity. Nevertheless, in 1970 the inter-regional bulk transmission and the final distribution of gas were together still costing the industry on average about 2.75 pence per therm.

The other secondary energy producer, the electricity industry, has also been able to take advantage of the falling real costs of the primary energy industries. In 1968 the current value fuel costs of the Central Electricity Generating Board (CEGB) per kWh sold were slightly below those of a decade earlier. In part this economy was also the result of the industry being able to exploit more efficient and cost-reducing technologies in both conventional and nuclear generation. The development of larger generating sets, from between 30 and 60 MW each in the late nineteen-forties, to 660 MW in the most recent stations of the CEGB, is but one example; the use of higher temperatures and pressures is another. Technical developments such as these succeeded in reducing the capital charges per kW of installed capacity from about £57 to £40 during the same period. The same innovations also offered economies in running costs by raising the efficiency of the new plants from approximately 31 to 38 per cent. In addition, whatever the disappointments, and the over-optimism of the apologists for nuclear power, it cannot be denied that both the capital and the operating costs of this new technology have also been substantially reduced (see Table 6.3). The first Magnox reactor at Berkeley, for example, has a generating cost of 0.51 pence per kWh, whereas the Oldbury plant which was commissioned five years later in 1967 has a cost of 0.29 pence per kWh. (The last station in the Magnox programme will undoubtedly generate at costs somewhat above the lower figure as a consequence of commissioning delays and inflation.) In addition to these advances in its production phase, the electricity industry has also been able to reduce its unit transport costs through the construction of its 400 kV super-grid to supplement the older 275 kV system, and through the development of techniques for the more efficient use of the grid system.

The coal industry has naturally not been able to leave these challenges unanswered. The introduction of power loading – which by 1970 represented nearly 95 per cent of the National Coal Board (NCB) output – and of other new techniques in the mining and preparation of coal have substantially improved the productivity of the industry. Output per man-year in 1969/70 was 464 tons; in 1960 it had been only 310 tons. In the meantime, the Board has concentrated production in its larger and more efficient collieries, reducing the number of pits to about 300 in 1970, compared with the 700 which were being exploited a decade earlier. Simultaneously, the workforce of the industry has been halved to some 300,000 men. The competitive stance of the coal industry has also been assisted by the government in its 1968 decision to write off £415 million of unproductive assets – a sum equivalent to £1,000 for every miner employed in the middle nineteen-sixties – and by improvements in the efficiency of coal transport and distribution. The introduction of merry-go-round trains with a payload of 1,000 tons reduced substantially the freight rates on coal hauled between at least some mines and power stations, a saving over the longer hauls of as much as 50 per cent. Similarly, the increasing movement of train rather than waggon loads of coal – the latter made up 60 per cent of all the coal handled by the railways at the time of the Beeching Report (British Railways Board, 1963) – has offered substantial savings. In addition, the concentration of the industry's distribution network upon fewer but larger depots has afforded yet further economies. The first of these, at Palace Gates, was designed to receive its coal from only five collieries and has replaced eight depots handling the coal of 35 mines; by 1970, the postwar scatter of some 6,000 distribution depots had been reduced to about 400.

No statistical series is available to demonstrate the changing costs of (delivered) energy in the British market. Nevertheless, all the available evidence suggests conclusively that – largely as a consequence of these many improvements in the efficiency of energy production and supply, but assisted by a persistent weakness in the international price of oil – the real costs of energy in Britain have fallen over the last ten or fifteeen years. Particularly has this been the case at the larger markets and in the more accessible locations. It is the purpose of this chapter, therefore, after examining certain associated responses by the several producer–supplier industries competing in the market, to explore some of the implications of this situation for public policy, with particular reference to some of the relatively neglected spatial aspects. The government's concern with the nation's energy economy is, of course, inescapable. Quite apart from the fact that, with the exception of the oil industry and most of the natural gas and oil exploration and production in the North Sea, the energy sector of the United Kingdom economy is under public ownership, the very size and economic importance of the energy industries make it impossible for any government to stand aside and ignore developments. In 1965 it was estimated that the energy industries employed about 1 million men or 4 per cent of the working population, that their capital investment of £1,000 million was approximately one-sixth of the country's annual total investment, and that the value of their net output was about five per cent of the national product (Ministry of Power, 1965, 1).

The effects of increasing competition

Associated with the falling real costs of energy provision in Britain during the last decade or so have been four noteworthy developments. These have been the emergence of keener market pricing, especially in the industrial fuel market; an accelerating rate of substitution of one fuel for another; a new awareness by the producer–supplier industries in the market of their total system costs; and the withdrawal of the higher cost and less profitable elements in each of the production–supply systems. Each of these changes is examined in turn.

Prices

From the middle nineteen-fifties onwards, the oil industry began seriously to challenge the dominance of the coal industry in the British energy market in general, and in the industrial fuel market in particular; by the late nineteen-sixties, it stood as a highly competitive force. The average price of energy in this sector *circa* 1968 stood at some 2.5 pence per therm delivered. But where an individual consumer was able and willing to negotiate a large contract over a number of years the price was lower, and was tending to fall. In 1969, for example, a group of oil companies (Shell-Mex and British Petroleum, Esso, Texaco, Phillips and Occidental) was able to secure a major contract with the British Steel Corporation for the supply of 4.3 million tons of fuel oil over three years. The £120 million agreement involved delivered prices which were reported to range from £5.40 to £6.90 per ton, excluding taxes and the Suez surcharge; with these charges, the contract worked out at an average delivered price of 2.1 pence per therm, and reflected in part the outstandingly flexible pricing opportunities of the oil industry, which can boast about 40 per cent of its sales in non-competitive markets.

At this time (1969) the gas industry was armed with the prospect of considerable supplies of relatively low cost North Sea natural gas, and was presenting a new and powerful challenge in the industrial fuel market. It had already signed a major twenty-five-year contract valued at £250 million with Imperial Chemical Industries, under which gas will be used partly as a naphtha replacement and partly for steam-raising and general heating purposes. The reported delivered price was 1.7 pence per therm. Although the delivered price was naturally higher for smaller quantities of gas and shorter delivery periods, many other new contracts were also secured by the gas industry in the late nineteen-sixties as it fought for and won a larger share of the national energy market, sometimes displacing oil but more frequently substituting for coal.

To such commercial aggressiveness, the National Coal Board was able to make the occasional dramatic response. In 1968, for example, it was able to secure a major contract with Associated Portland Cement for the sale of about 1 million tons of coal per year at a delivered price of less than 1.7 pence per therm. This price was probably negotiated, however, more with an appreciation of the industry's considerable surface stocks of coal which at the time exceeded 30 million tons than in relation to its contemporary mining and transport costs. The even lower price of just over 1 penny per therm which was quoted for the sale of coal to the Alcan smelter in Northumberland is, of course, a pithead price to which a small transport charge has to be added for comparative purposes. The negotiation took place in the face of severe competition from the electricity industry, which was subsequently able to conclude agreements with the other two aluminium smelting companies at Anglesey and Invergordon; the financial arrangements associated with these twenty-five-year contracts were highly complex and involved both capital payments to the CEGB and the North of Scotland Hydro-Electric Board plus generating charges based upon hypothetical costs for the advanced gas-cooled reactor.

Substitution

The ability and/or willingness of the various energy industries to hold or even reduce their prices has naturally varied through time and space – just as have improvements in their efficiency. This feature has expressed itself most forcibly and predictably in the steady substitution of one fuel for another. Whereas in 1969 some 51 per cent of the country's energy needs were satisfied by coal, ten years earlier the percentage had been 77; during the same decade, the oil industry increased its share of the national energy market from 22 to 43 per cent (Table 6.1). Part of this shift in energy demands was, of course, a function of the changing structure of the economy and especially the rapid growth of the non-competitive markets for oil such as motor and aviation spirits. Differences in the regional structure of the country's economy also accounted in part for the greater relative importance of, say, petroleum in the South East and South West

TABLE 6.1. *United Kingdom: inland energy consumption,*
1949, 1959 and 1969 (million tons coal equivalent)

	1949		1959		1969	
	m.t.c.e.	%	m.t.c.e.	%	m.t.c.e.	%
Coal	199.0	90.4	192.4	76.5	163.7	50.7
Oil	20.2	9.2	57.0	22.6	137.9	42.7
Natural gas and colliery methane	—	—	0.1	0.1	8.5	2.7
Nuclear electricity	—	—	0.5	0.2	10.7	3.3
Hydro-electricity	0.8	0.4	1.5	0.6	2.0	0.6
Total	220.0	100.0	251.6	100.0	322.8	100.0

Source: Ministry of Technology, *Digest of Energy Statistics 1970.*

regions than in the Northern and the Yorkshire and Humberside regions. Although regional data on final energy consumption by primary source are not available – in part as a consequence of the complexity of the inter-regional transfers of secondary energy, especially electricity – it is clear that the rate of substitution of one source for another in the past has exhibited significant spatial variations, a point to which further attention is given later in this chapter.

System costs

As competition has stiffened and as many prices have fallen in real terms during recent years, the profit margins of the energy industries have tended to narrow. Consequently, each of the industries has had its attention turned more frequently to the expenses involved both in operating and in adding to its facilities; and as a result each has become increasingly aware of its total system costs. Attempts to contain and if possible reduce these costs have become increasingly searching and sophisticated in each of the producer–supplier industries. Since April 1967, for example, the Gas Council has invested heavily in an outstandingly good operations research team in an attempt to minimise their capital investment and total system costs during the critical period of conversion to natural gas. Assuming that there will be supplies of and a market for gas ranging between 57 and 114 million cubic metres (2,000 to 4,000 million cubic feet) per day by 1975, the team developed a 'gambit model' of the gas economy and sought to optimise the size, the location and the timing of the laying of pipelines, the construction of storage facilities and the installation of compressors. The objective was to provide the industry with an investment strategy which would minimise (the present value of) its production and transmission costs in the early and middle nineteen-seventies (National Board for Prices and Incomes, 1969). The electricity generating and transmission industry has similarly become increasingly concerned with the quality of its investment decisions. Every major new project of the CEGB, be it a new power station or an extension to the 275 kV or 400 kV grids, has therefore come to be assessed in terms of its effects upon the Board's total system costs, and internal rates of return are calculated for any new capital which might be invested and for alternative strategies which might present themselves. In this fashion, alternative proposals – such as the competing claims of coal-fired and nuclear power stations for incremental generating capacity – can be assessed to determine which investments appear most likely to minimise the (net present) cost of a modified production and transfer system. The older and simpler type of Weberian analysis which formerly helped to determine the spatial pattern of power station and transmission line investment (Rawstron, 1955; Manners, 1964) has therefore given way to a much more thorough and quantitative systems appraisal of alternative development opportunities which might lie before the industry (Berrie, 1967; Hauser, 1969).

The oil industry, both nationally and internationally, has been one of the pioneers in the application and development of operations research, exploiting the discipline's techniques to find – within the industry's distinctive institutional framework – 'optimum' solutions to its highly complex refining and transport problems. But the coal industry is different and, compared with its competitors, it has been less able to exploit the many new management and investment appraisal techniques. Whilst it has been in a position to discover and exploit many new cost-saving procedures for the production and preparation of coal, it is only in the case of coal transfers to the power stations of the CEGB that the Coal Board has been able to 'optimise' a significant part of its transport operations through the use of linear programming techniques. For, in strong contrast to the other energy industries, the NCB has only a limited degree of control over much of the coal transfer and distribution system. The contrast is important.

Whilst all the producers and suppliers of energy in the British market serve from a limited number of production points demands which have a distinct spatial expression, the transfer and distribution arrangements of each type of energy – in terms of its reliance upon particular modes of transport, in its institutional arrangements and hence in its adaptability to advanced management procedures – are characterised by important differences. The nationalised electricity and gas industries, for example, are engaged in the production, transfer and distribution of energy almost entirely in and through their own facilities. Theoretically therefore they have an almost totally integrated production–supply system. Certain institutional barriers naturally create some divergence between theory and reality. In the case of the electricity industry, the CEGB is responsible for the production and bulk transmission of electrical energy in England and Wales, but Area Boards are concerned with its distribution and sale; such a division of responsibility naturally limits to some degree the ability, if not the desire, of the managements concerned to optimise the performance of the whole system. In Scotland, there are two separate authorities – the North of Scotland Hydro-Electric Board and the South of Scotland Electricity Board – which are responsible for the generation and distribution of electricity in the country. Similarly, in the case of the gas industry, management and accountability are divided between the Gas Council, which is responsible for the purchase and bulk transmission of North Sea and other (imported) methane gas to all parts of Britain, and the Area Boards which handle its distribution and sales as well as continuing to produce secondary gas from coal and oil feedstocks. However, the electricity and gas industries each use only one mode of transport to transfer their energy, and the opportunity exists for the managements particularly concerned with different phases of the industry and its different spatial components to forge close links and enjoy close cooperation. As a result, efficiencies and economies are potentially available to these two energy industries which are not always shared by their competitors.

The oil industry, for example, makes use of a much greater variety of transport modes – ocean and coastal tankers, barges, railways, pipelines and road haulage. Moreover, it is part of an international industry and is structured within an oligopolistic, and to some degree competitive, framework. Although this structure may impose upon the industry certain opportunity costs (of scale economies foregone, of unnecessary cross-haulage of products and of over-investment in retail outlets), there can be little doubt that the competitive stance of the industry *vis-à-vis* other sources of energy is considerably enhanced by the economic benefits which stem from its highly integrated set of production and supply systems. The vertically integrated international companies which dominate the industry handle the oil from the production wells to the market outlets, and only in the case of rail transport, plus a small percentage of the road haulage fleet, are the facilities not owned by the industry. Even in the case of oil transfers by rail, the tank cars are invariably owned by the oil companies, and any freight rate negotiations are influenced by the possibility of further pipeline construction. At the final distribution phase of the industry, where the major oil producers and refiners share a small part of their activities with merchant distributors, both the style and the quality of the service offered by the latter is decisively influenced by standards set by the international industry.

The coal industry, however, stands in quite a different position. Despite the nationalisation of its mining and preparation phase, the industry is characterised by a highly fragmented production-and-supply system. Much of the transport of coal from the pithead to the largest markets and to the industry's distribution centres is outside the direct control of the NCB. The transfer is achieved mainly by rail, to a lesser degree by road and to an even smaller and declining extent by water transport operations. The rail waggons are in part owned by the Coal Board, which has also built up its own fleet of lorries. But in all other respects the transport services used by the coal industry are bought from British Rail, from private road hauliers and from coastal carriers. The near monopoly position of British Rail over a large proportion of the country's coal transfers from the pits towards the markets – the railways handled nearly 70 per cent of all movements in 1969 – plus the lack of innovation in this facet of their operations has had two serious consequences. The rationalisation of coal transfer operations has been slow and on too small a scale; and coal freight rates have not only remained relatively high per ton-kilometre, but they have also moved steadily upwards. The competitive position of the coal industry has suffered badly as a consequence. Innovation and the rationalisation of coal movements by rail have only been forthcoming where short- and medium-haul road services have challenged for the traffic, in situations where the adoption of the experimental technology of transporting coal through pipelines appeared imminent and, in greatest measure, where outstandingly fierce competition from other fuels in major energy markets threatened to eliminate coal and hence significantly to reduce an outstanding source of railway revenue; the Generating Board's power stations were a case in point. In situations such as these, selected rail rates have been reduced, some of them significantly. However, the overall situation continues to compare poorly with the efficiency of rail-borne movements of oil.

In addition, the final distribution of a considerable share of coal sales is in the hands of a large number of wholesalers and merchants over whom the NCB has traditionally had very little control. Their size, their efficiency, the quality of their management and their ability to raise capital vary enormously. The Board has therefore sought in recent years to exert an increasing influence over this situation by injecting some of its own capital into the modernisation and mechanisation of large depots, and by trying to improve the standard of the industry's distribution services by appointing 'Approved Coal Merchants' who are able to meet certain efficiency criteria. But the size of the task is enormous – the industry includes 12,000 retail merchants in addition to the wholesalers and the Coal Board's own sales organisation – and progress has been predictably slow. Thus, although the NCB has recognised the weakness of the coal industry's transport and distribution phase, and has helped to achieve some improvements in its quality and efficiency, the industry nevertheless remains burdened by an essentially fragmented production and supply system. Whilst there have been notable achievements in improving the efficiency of the production phase of the coal industry – and, indeed, plenty of opportunities still remain – this fragmentation is particularly burdensome when market pressures increasingly demand the reduction of the industry's total system costs.

Elimination of marginal system elements

The final noteworthy development in the British energy market in recent years, and stemming from the intensification of inter-fuel competition, is complementary to this new awareness of total system costs. It is the growing willingness of the energy industries to eliminate the higher cost and less profitable elements in their production–supply systems. The most dramatic example of this has been the mine closure programme of the NCB which, reaching its peak during the years 1966–9, had a distinctive geography and ensured a rapid contraction of production in South Wales, Lancashire, Scotland and North-East England; the lower cost East Midlands and South Yorkshire mining areas were left in a relatively more important position. But the coal industry is not unique in this matter. During the nineteen-sixties, the gas industry virtually completed its programme of closing down many small and high cost works (Manners, 1959), and the electricity Generating Board phased out some of its older and less efficient plants. Rationalisation programmes to modernise the distribution arrangements of both the oil and coal industries served to eliminate those depots with higher unit costs. Although each of the energy industries continued to claim the availability of their products on a nationwide basis, it is clear that in some areas and for some fuels this was only at a price which consumers would be increasingly reluctant to pay. In the 1965 White Paper on *Fuel Policy* (Ministry of Power), in fact, the government formally removed the obligation on the gas industry to maintain indefinitely their unprofitable services in rural areas, in small country towns and in other small markets, exemplified by Mid-Wales, much of the south west of England and large areas of Scotland. The government took the view that there was little reason for both the publicly owned gas and electricity industries to continue to subsidise energy supplies in the same localities.

In this context, however, it should be noted that the profitability of supplying energy to a particular market does not lend itself to simple calculation, for each of the producer–supplier industries is characterised by joint spatial (as well as functional) costs, and they willingly practise a considerable degree of spatial cross-subsidy. The most overt, and extreme, case of spatial cross-subsidy practised in the production phase of the energy industries in Britain is that of the NCB. Throughout most of the post-war period, its objective has been broadly to finance the capital costs of all its production units from the profits of its East Midlands and South Yorkshire pits; the other coalfields were given a more limited target of covering only their operating costs, a target which they only occasionally surpassed. Something of the magnitude of the variations in the profitability of the several British coalfields is indicated in Table 6.2, which records the gross profit situation of the eighteen NCB areas; from these figures should be deducted a charge for interest, which on average amounted to 23.5 pence per ton of saleable coal for all colliery areas. Some of the dilemmas which this policy presents in the matter of investment resource allocation have been discussed by Shepherd (1964). Since 1962, but more especially since 1966, there has been a growing insistence by the NCB not only upon coalfield but also area accountability, and inter-coalfield differentials in the pithead price of industrial coals have been established and widened. Nevertheless, a substantial degree of spatial cross-subsidy remains as a dominant feature of the economic geography of British coal production.

The coal industry, however, is unique only in the degree to which it behaves in this way. All the other energy producers, with zone-delivered prices for at least some categories of their sales, serve their markets from facilities which exhibit a considerable range of costs and which include plants of doubtful profitability. In the case of the electricity generating industry, for example, the unit costs of a large modern plant on base load are much less than those of the small, older units which are reserved for peak generation and stand-by duties; yet, in the absence of pricing policies which are designed fully to reflect these peak load costs, it would appear very unlikely that the latters' expenses are in fact covered by the revenue

TABLE 6.2. *National Coal Board Areas: costs and profits,*
1967/8, 1968/9 and 1969/70 (per ton of saleable coal)

Area	Num-ber of mines at end-year	Output of saleable coal (million tons)	1967/8 Costs (£)	1967/8 Gross profit (pence)	1968/9 Costs (£)	1968/9 Gross profit (pence)	1969/70 Costs (£)	1969/70 Gross profit (pence)
Scottish North	10	5.06	5.63	−48.4	5.74	−41.5	6.09	−52.1
Scottish South	22	6.49	5.21	7.4	5.19	20.9	5.92	—17.6
Northumberland	16	6.96	4.69	14.8	4.74	17.6	5.03	1.2
North Durham	19	5.44	5.34	9.1	5.56	−13.1	6.31	−73.8
South Durham	15	8.42	5.04	16.4	5.37	−13.1	5.63	−10.6
North Yorkshire	22	9.36	4.32	0.0	4.27	− 0.8	4.56	−29.5
Doncaster	11	8.66	4.50	−7.8	4.26	11.9	4.91	−34.1
Barnsley	23	8.19	4.57	0.8	4.66	−68.5	5.21	−56.6
South Yorkshire	19	10.27	4.42	45.5	4.31	46.4	4.84	21.7
North Western	17	7.12	5.76	−5.7	5.65	16.4	6.11	0.0
North Derbyshire	15	10.76	3.90	24.6	4.01	0.4	4.24	− 0.4
North Nottingham	15	11.96	3.83	43.9	3.77	37.3	4.01	38.6
South Nottingham	12	11.32	4.02	−8.2	4.03	3.6	4.08	20.5
South Midlands	15	9.40	3.80	24.2	3.96	10.2	4.19	14.0
Staffordshire	13	8.69	4.60	46.0	4.67	39.4	4.81	49.6
East Wales	30	7.84	5.69	49.6	6.15	9.8	6.79	−26.3
West Wales	22	5.15	6.57	−17.2	6.80	−32.0	7.38	−63.2
Kent	3	1.11	6.16	−96.7	5.97	−65.2	7.57	−211.6
Total	299	142.28	4.72	13.5	4.75	8.2	5.12	8.2

Notes: Excludes opencast production, which amounted to 6.68 million tons in 1969/70.

Gross profit is before charging interest, which on average amounted to 20.5 pence per ton of saleable coal for all colliery areas in 1967/8, 22.7 pence in 1968/9, and 23.5 pence in 1969/70.

Source: National Coal Board Annual Reports and Accounts.

they produce. Again, in the transport of energy, and especially in the case of an electricity grid or a pipeline network, facilities are shared between regions and areas in a fashion which makes it exceedingly difficult, if not impossible, to allocate costs precisely between them. The large diameter natural gas pipeline from Bacton to the Midlands and South Wales, for example, passes through or near to a number of small urban and industrial centres which in themselves could not possibly justify the installation of such a major transmission facility with its low unit costs of transport. Consumers in these smaller markets, therefore, although their demands possibly (but only marginally) improve the economics of the pipeline as a result of the small increments they may make to its load, benefit considerably from a transport facility which is justified primarily by the large urban and industrial demands in other parts of the country. Further, in the distribution and marketing of energy, the adoption of a zone-delivered pricing policy by definition shifts some of the costs of supplying the relatively small consumer in a comparatively remote location onto the larger consumers near to the points of energy production or bulk reception.

Although it is easy to point to the existence of, say, spatial cross-subsidies in the supply of electricity to the smaller communities of West Cornwall and Westmorland, or to the existence of subventions to the motorist buying petrol in the Highlands of Scotland, it is extremely difficult to quantify their magnitude. Indeed, with certain large costs in the production–supply system being incapable of accurate spatial allocation, it would be quite inappropriate to do so. Thus, the downward pressures upon real prices and the narrowing profit margins in the market for energy are inclined initially to generate further spatial cross subsidies in the system; and the closure of the more expensive production facilities, and/or the withdrawal of some suppliers from their higher cost and possibly profitless markets tend to be deferred as long as possible. However, this is a response which cannot continue indefinitely.

Energy costs in the nineteen-seventies

Before turning to some of the public policy implications of these recent and contemporary features of the British energy market, it is important to establish whether they are likely to persist into the forseeable future. An examination of the best evidence available on the prospective cost position of each producer-supplier suggests that the outstanding characteristics of the market over the last decade – the downward trend in real costs and an associated intensification of inter-fuel competition – will continue for the foreseeable future. There are, however, some important contrasts between the energy industries. Recent (1969/71) developments in the international oil market suggest a strengthening of those forces which are likely to encourage a hardening of crude-oil prices. The temporary closure of the Trans-Arabian pipeline, a series of major tanker disasters and an unexpected upsurge of global oil demand were of course essentially short-term influences upon price. But the continuing efforts of the Organisation of

Petroleum Exporting Countries to increase the f.o.b. return to the oil producing countries are more permanent; also they are bearing some fruit, as witness the 1970/1 acceptance by the international oil industry of higher posted prices on Libyan and Persian Gulf crudes, and the implied shift in the spatial incidence of profits. In addition, the re-evaluation of the role of oil import quotas into the United States – quotas which have been out-standingly influential in keeping down the delivered price of oil in the Western European market (Schurr, 1960) – could possibly lead to a sig-nificant short-term upsurge in the demand for Middle Eastern and African crudes. However, in the slightly longer term, the ease and economy with which supplies (from those same sources) can be increased, plus the dis-covery of large and significant finds in Alaska and the North Sea, together seem likely to push real prices in the opposite direction. As a consequence, it remains difficult to dissent from Adelman's (1964a; 1964b) thesis that for a number of years to come more oil will be looking for markets than there will be markets available. The inevitable result will therefore be a continuing downward drift in the real price of oil f.o.b. With this tendency must be associated the prospect of larger supertankers and an increasing proportion of British imports being carried in them; larger refineries with their economies of scale; larger coastal tankers, rail cars and road tankers to distribute the products; and yet further improvements in the efficiency of the industry's distribution arrangements. The real cost savings potentially available over the next decade to the oil industry, both internationally and in Britain, are therefore considerable. Armed with a flexible pricing system and with a captive market for about 40 per cent of its sales, the oil industry is clearly in a strong position to offer and to meet competition in the British energy market of the nineteen-seventies.

The gas industry can also look forward to the prospect of further improve-ments in its efficiency and a continuation of its recent real cost reductions. The Gas Council is in the singularly advantageous position of having recently negotiated twenty-five-year contracts equivalent to 40 million tons of coal per year. Some of these contracts specify a slight reduction in the beach price of the gas through time, but inflation is likely to be much more effective in reducing the real cost of the fuel. Assessments of the gas industry's market prospects vary (Odell, 1966). Short of entering the base load electricity generating market, however, it would appear that the gas industry is likely to have to employ vigorous salesmanship and to adopt competitive pricing policies if it is to quadruple the industry's 1969 level of sales in a matter of five or six years and so be able to take all the con-tracted North Sea gas. In such a circumstance, it is most unlikely that any further contracts will be negotiated with the explorer–producer companies at a significantly higher beach price. Moreover, once the major investments in a new transmission and improved distribution system have been completed by the early nineteen-seventies, the gas industry will be able to make incremental changes to it – through the addition of further compressors, new storage facilities and the like – at relatively small cost; as a result, real transport costs per cubic metre will tend to fall.

The electricity industry also can look forward in the medium term to real cost reductions. With the recent availability of natural gas to provide a cheap source of fuel for summer generation (when the demand for gas is relatively small), its primary energy sources have now been increased to four, and price competition between the three conventional fuels is unlikely to abate. The industry's generating sets and stations will continue to get larger. The full impact of the 500 MW sets on the industry's costs has yet to be felt (except in so far as teething troubles and break-downs have inconvenienced the CEGB); a 660 MW plant is to be installed in the near future for the first time and 1,300 MW sets are now being designed. One of the major uncertainties which the electricity industry faces is, of course, the economics of nuclear power, a matter of singular complexity and controversy. However, as in the recent past, two developments seem highly probable. First, the (real) unit capital costs of nuclear power stations will continue to fall. Second, since capital costs dominate the total costs of nuclear stations, the costs of generation also will fall (Table 6.3). That nuclear power still remains more expensive than conventional generation using untaxed oil is agreed under the most commonly used evaluation assumptions; pre-tax generation costs at Pembroke, to be commissioned in 1971 and burning oil, are forecast at 0.18 pence per kWh. Nevertheless, hopes remain high and not unreasonable that the later stations of the second nuclear programme – and certainly the fast breeder reactors to follow – will offer even cheaper base load electricity.

The prospects for coal production are much more difficult to assess. The former Chairman of the NCB, Lord Robens, sought to assure the industry and indeed the public that there are enormous opportunities for cost reductions. Asserting that annual increases in productivity of 9 and 10 per cent are within the production industry's grasp, he has pointed to the fact that British mining machinery in the mines of the United States yields four times the tonnage per manshift compared with its performance in the better British mines (Robens, 1968). With 'intelligence, information and cooperation' (Robens, 1970), it is probably true that the industry is capable of major cost reductions. If the absentee rate of the industry were

TABLE 6.3. *Great Britain: trends in nuclear power costs*

Station	Date commissioned or to be commissioned	Station size (MW)	Capital cost (£/kW)	Efficiency (%)	Generation costs (pence per unit sent out)	
					(a)	(b)
Magnox stations						
Berkeley	1962	276	185	24.4	0.51	0.39
Bradwell	1962	300	175	28.2	0.45	0.36
Hunterston 'A'	1964	320	176	28.3	0.44	0.36
Hinkley Point 'A'	1965	500	150	26.4	0.40	0.35
Trawsfynydd	1965	500	143	29.4	0.40	0.32
Dungeness 'A'	1965	550	116	32.9	0.30	0.25
Sizewell 'A'	1965	580	107	30.5	0.29	0.25
Oldbury 'A'	1967	600	113	33.6	0.29	0.24
Wylfa	1971	1,180	111	—	—	—
AGR stations					(c)	(d)
Dungeness 'B'	1972	1,200	105	41.5	0.21	0.24
Hinkley Point 'B'	1972	1,300	76	41.7	0.20	0.23
Hunterston 'B'	1973	1,300	67	41.7	0.19	0.22
Hartlepool	1974	1,300	71	—	—	—
Heysham	—	2,500	—	—	—	—
Sizewell 'B'	—	2,500	—	—	—	—
Oldbury 'B'	—	650	—	—	—	—
Portskewett	—	1,300	—	—	—	—
Connah's Quay	—	2,500	—	—	—	—
Stake Ness	—	1,250	—	—	—	—

Notes: Assumptions underlying generating costs are:

	(a)	(b)	(c)	(d)
Interest (%)	$7\frac{1}{2}$	$7\frac{1}{2}$	$7\frac{1}{2}$	8
Load factor (%)	75	85	75	75
Life (years)	20	25	20	20
Fuel burn-up (MW days/ton)	3,000	4,000	18–20,000	18–20,000

Capital costs include an initial fuel charge; costs are not adjusted for inflation.

Connah's Quay and Stake Ness (which might be a heavy water reactor) still awaited ministerial approval in November 1970.

— Information not available or not known.

Sources: Central Electricity Generating Board; South of Scotland Electricity Board; Atomic Energy Authority.

to be reduced from its present 18 per cent to a more acceptable 6 or 8 per cent, once again major reductions in production costs per ton would be forthcoming. Yet, for reasons rooted firmly in its history, it is doubtful whether the industry is capable of taking advantage of these opportunities. The coal industry's wage costs, on the other hand, are certain to increase. After a substantial wage award in 1969, a further large claim unrelated to productivity increases was partly conceded in 1970. In the transport and distribution phase of the industry, opportunities undoubtedly exist for the introduction of larger (possibly unit) trains and for the bargaining of lower rates. The further mechanisation of handling at the larger depots and the rationalisation of distribution arrangements could be pushed forward more vigorously. All such developments will afford important economies in real terms. However, the fragmented nature of the production–supply system once again suggests that full advantage is unlikely to be taken of all the opportunities available.

An hypothesis

The prospect of the British energy market remaining highly competitive – and perhaps becoming increasingly competitive – in the nineteen-seventies prompts speculation in the following general terms. With the market prices for many forms of energy tending to fall in real terms, the probability exists that, in some industries and in some places, prices will fall faster than costs; profit margins will narrow; and occasionally profits, by whatever yardstick they are measured, will disappear. As a result, there could be further pressure upon particular producer–supplier industries to withdraw from particular markets.

In this connection, it is known that each of the energy industries has a distinctive geography of costs. The production expenses of the coal industry are lowest in the East Midlands. There, the Coal Board's South Midlands division has achieved a level of productivity which is half as much again as the national average – three tons per manshift compared with the national two tons. There, too, it has been claimed by the industry's spokesmen the marginal cost of production in 1968 was as low as 1.20 pence per therm. Wage inflation is likely to have raised this figure to 1.35 pence by 1970. However, given the relatively high costs of moving coal even by the most efficient train operations, it is clear that the delivered costs of the industry's most competitive source of energy rise steeply from this East Midlands' centre. Isodapanes can be drawn upon the map of Britain to represent in a crude fashion this spatial variation in the delivered costs of

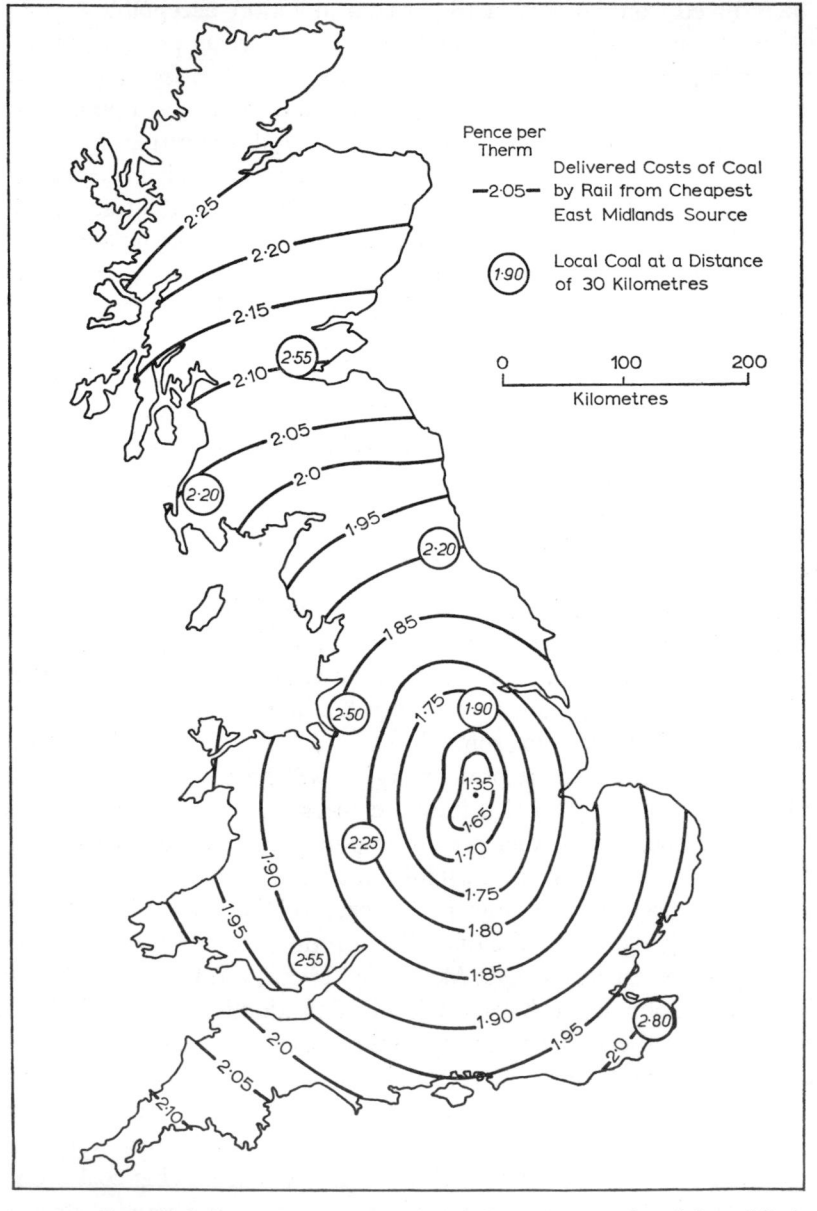

Figure 6.1. Great Britain: estimated marginal delivered costs of coal from National Coal Board mines, for large consumers, *circa* 1970.
Source: Hauser, 1969, 309 and author's own estimates.

coal from the East Midlands (Figure 6.1). This map suggests that in the early nineteen-seventies the delivered cost of a marginal ton of coal in the larger markets is likely to be in the order of 1.90 pence per therm in London, and less than the cost of coal produced locally in all the other coalfields, if we assume a short road or rail haul of say 30 kilometres on the locally produced coal; in the more peripheral markets, the industry's marginal costs are likely to exceed 2.00 pence per therm. Such considerable variations in the costs of delivered coal are already partly reflected in the pricing policies of the Board. In 1968, for example, the price of household coal in Aberdeen was 36 per cent higher than in Nottingham – just under £15 per ton compared with just under £11 per ton. This is a spatial price variation unmatched by any other energy industry, as will be seen.

The lowest cost points for natural gas are on the east coast, at Bacton, Mablethorpe and Easington, where the North Sea pipelines come ashore (Figure 6.2). Costs then vary spatially with the distance that the gas is pumped through the transmission system, so that the off-take points with the highest costs to the industry (but not of course the highest prices, since these are spatially uniform in response to public policy) are in the south west of England and in Scotland. When large diameter, high pressure pipelines are used with high load factors, the cost of transport does not rise particularly steeply with increasing distance – between 0.20 and 0.30 pence per therm may be added to the beach cost of 0.85–1.10 pence per therm by the time the gas has reached these more distant markets. Small diameter pipes and deteriorating load factors, on the other hand, raise the cost of gas transport considerably. Therefore, it is reasonable to expect the delivered costs of natural gas to vary much more intra-regionally than inter-regionally, with the unit costs of satisfying small markets in East Anglia and Lincolnshire standing appreciably above those of large industrial consumers in Greater London or Merseyside.

In contrast to the two other primary energy industries, the oil industry has a much more uniform geography of costs. There are small differences in the delivered costs of crude oil, depending partly upon the distance from and the nature of the oil fields being used; partly upon the size of tanker being used, and the charter arrangements which have been agreed; and partly, since the industry is integrated both vertically and internationally, upon the advice of the industry's accountants concerning the most appropriate place for profits to be accumulated. Clearly, the refineries on Milford Haven and at Fawley are likely to have marginally cheaper crude oil than the refineries at Llandarcy (Swansea) and Grangemouth, both of which have the costs of a pipeline connection with deepwater berths at

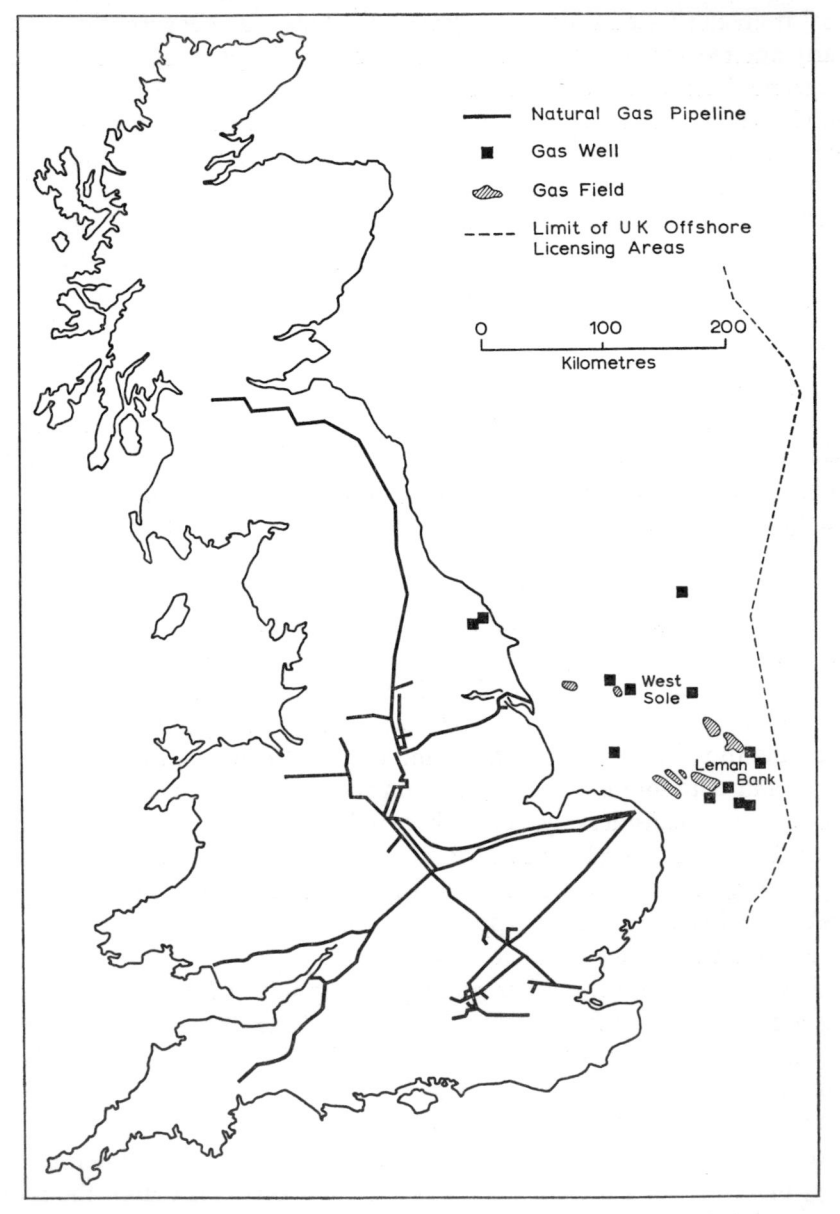

Figure 6.2. Great Britain: natural gas fields and primary transmission system, existing and planned, 1970.
Source: the Gas Council.

Milford and Finnart respectively. Yet a further influence upon the variations in the oil industry's spatial costs is the scale of refinery operations; *ceteris paribus*, the lowest cost centres of oil product manufacture in Britain would be at the large refineries on Southampton Water, lower Thamesside and the Mersey.

Although these refineries are concentrated in the southern part of the country, it is equally important to note that the oil industry also has a scatter of rather smaller refineries well distributed around the shores of Britain – in South Wales especially at Milford Haven, on the Forth and (prospectively) on the Clyde, and on both Teesside and Humberside (Figure 6.3). Their refining costs are naturally somewhat above those of the country's largest plants, but the differentials are steadily being eroded as their capacity and throughput are expanded. Well placed to serve the regional energy markets of the country, this geography of refinery capacity plus the highly efficient and the relatively low cost nature of oil product transport has one outstanding consequence; this is the comparatively small difference in the costs to the industry of importing, refining and distributing oil to almost any part of the country, whether it is Plymouth or Dundee, Birmingham or London. Data on the actual geography of the oil industry's costs are not available – and in any case their joint nature would make any single set of figures for, say, fuel oil highly suspect. However, it is noteworthy that there is less than a 5 per cent variation in gas–oil prices (and, indeed many other products) delivered to different parts of Britain – a figure which compares strikingly with the 36 per cent variation noted earlier for household coal. Figures compiled by Hauser (1969) show a variation of only 0.15 pence per therm in the estimated price of bulk fuel oil for the electricity generating industry at selected coastal points in the early nineteen-seventies (Figure 6.3). The range lies between 1.05 pence per therm (1.49 pence with tax) and 1.20 pence per therm (1.64 pence with tax). Even adding a transport charge for deliveries inland, this is a geography of price which reflects a geography of costs, and which clearly presents an enormous challenge to the other energy industries, the coal industry in particular. In discussions of inter-fuel competition, it is often neglected as a major facet of the oil industry's commercial strength.

With much the largest spatial variations in its costs, its prices and its profitability, in theory at least the coal industry cannot avoid being faced with increasing pressures to reconsider its geographical role in the British market. Especially where its costs can be clearly isolated and are high, the wisdom of continuing to supply all but any premium energy demands in those parts of the country which are furthest away from its low cost East

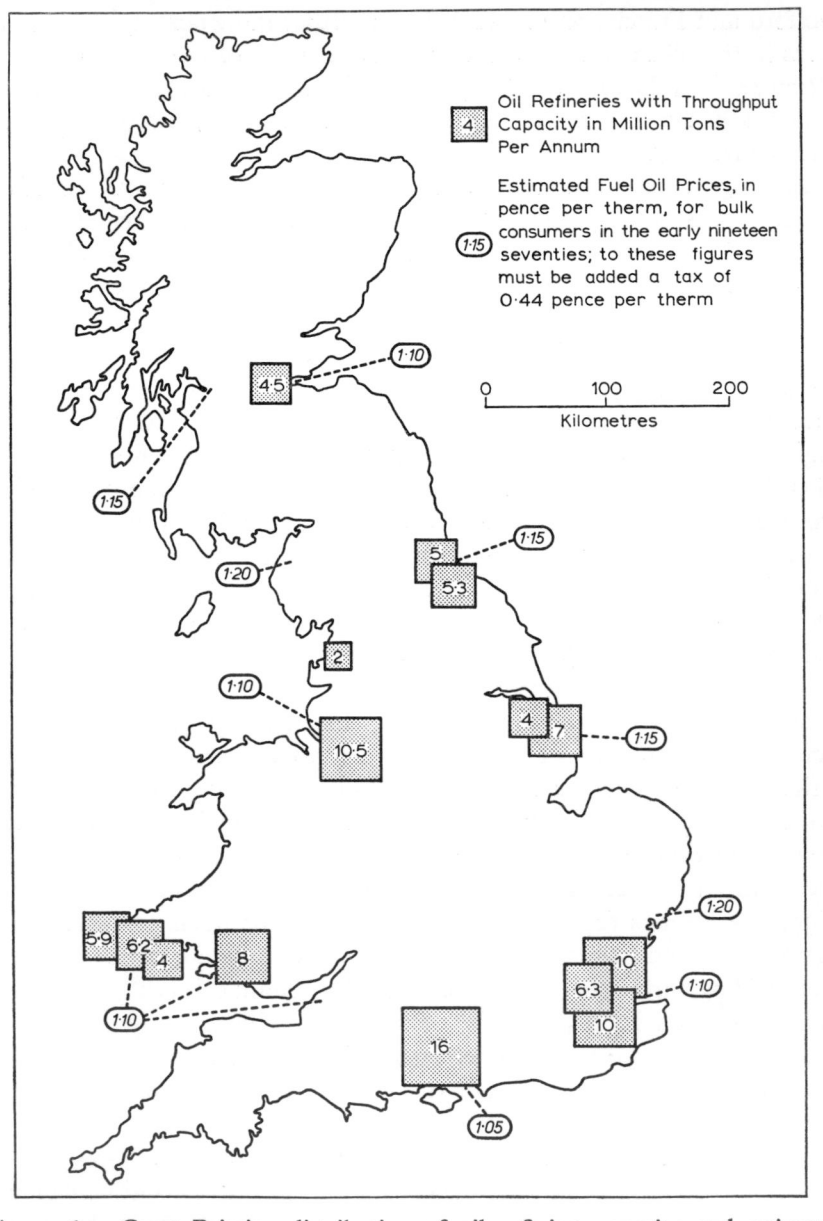

Figure 6.3. Great Britain: distribution of oil refining capacity and estimated delivered prices of fuel oil, *circa* 1970.

Sources: Ministry of Technology, 1970; Hauser, 1969; and author's estimates.

Midlands' base will have to be questioned and the hope of finding new ones abandoned. Naturally, its response will be through the introduction of further inter-regional price differentials rather than a formal withdrawal of supplies. Similar, although less urgent, pressures might reasonably be anticipated for the gas industry too in its highest cost regions. But these are considerations which are unlikely to affect the oil industry. However, theory and reality are liable to diverge as a result of three factors in particular. The first is the fact that in the short run many consumers are tied to a particular fuel through their energy-using equipment; demand as a consequence is somewhat inelastic, more especially in an economy the overall growth rate of which is somewhat low. The second, to which attention has already been drawn, is the complexity of the spatial cross-subsidies which characterise all of the energy industries, and which pose enormous difficulties for them in their sub-regional and local accounting. In the short run, once again, these will tend to reduce the effects which widening contrasts in the geography of energy costs might be expected to generate. The third is the commercial and financial behaviour, and the politics, of public enterprise.

Public attitudes towards the commercial and financial objectives of the energy industries in the public sector have evolved through time, as the older ideas of Morrisonian socialism have given way to state capitalism. During the nineteen-sixties, and more especially following the publication in 1967 of a White Paper on *Nationalised Industries – A Review of Economic and Financial Objectives* (Ministry of Technology), an attempt has been made to ensure that the financial objectives of the industries are specific and rigorous. Target rates of return upon capital, test rates of discount and more rigid constraints upon their ability to borrow capital were all laid down for each of the nationalised industries. In many respects public enterprise has been set very similar goals to the large corporations of the private sector of the economy – and certainly many of the investment decision-making procedures which they have recently embraced, such as the use of discounted cash flows and the calculation of internal rates of return upon additional capital investment, originate from the practice and experience of the private sector. However, some of the objectives set for public enterprise are clearly less realistic than others. Pricing policy is one; the ultimate financial discipline of some public industry is another.

In the 1967 White Paper, all the nationalised industries were enjoined to price their products at their long-run marginal costs. In the case of the energy industries, of course, such marginal costs have a distinctive spatial expression. It would clearly cost the Gas Council more to supply an extra million therms of gas at Aberystwyth than at Hull. Yet is it not at all certain that the industries are willing or able to reflect these cost differentials in their prices – for both theoretical and practical reasons. Theoretically, there are posed all the problems of allocating in space all the joint costs of the production–supply system. Practically, there is likely to be strong public opposition to a public industry which elected to publish substantially different tariffs for, say, the Highlands of Scotland and the Central Valley, or for Westmorland and south-east Lancashire. At both the inter-regional and intra-regional scales, all the questions of regional development policy are raised and compounded with issues of public relations. Even the privately-owned oil companies feel the need to subsidise sales in the remoter parts of Britain and are reluctant to charge a substantial supplement to their customers in Sutherland and Skye. The basic point to be made, however, is this: to the extent that public enterprise does not reflect in its pricing policies its long-run marginal costs, it will be less likely to meet its financial objectives.

Alongside these dilemmas of pricing, the ultimate financial discipline of public enterprise must also be questioned. Whilst the 1967 White Paper set out a rigorous basis for investment decision-taking, in the same year the 1967 Coal Industry Act was put on the statute book. This deferred until 1971 the obligation of the NCB to put its house in order, to compete freely in the primary fuel market for electricity generation and to meet its financial objectives. The Coal Industry Act of 1971 not only continues until 1974 the various government grants to assist the coal industry in its contraction and manpower redeployment; it also raises the statutory limit on the NCBs deficit from £50 to £75 million, with powers to increase it again within a ceiling of £100 million. In other words, it is perfectly clear that the full impact of increasing competition and falling real prices in the British energy market, more especially in its spatial expression, will be blunted by a number of factors, not least the financial indiscipline of public enterprise. Nevertheless, that impact cannot be ignored completely.

National energy policy

The relevance of this analysis to public policy stems especially from the various attempts which were made in the nineteen-sixties to formulate a national fuel policy. Before 1965, various *ad hoc* measures relating to particular energy industries were approved by successive governments – the long-standing ban upon coal imports (lifted in December 1970), and the investment of over £1,000 million of research funds into nuclear power are but two examples. Yet no attempt was made to espouse an overall and consistent view of the British energy economy until the Minister of Power published his White Paper on *Fuel Policy* (Cmnd. 2798) in 1965. For the first time in that document the government attempted to spell out a consistent attitude to the various energy industries and the evolution of the energy market. However, it was only to be an interim statement, in so far as the magnitude of the reserves of natural gas under the North Sea at that time remained uncertain. Two years later, therefore, in 1967 the government presented to Parliament a revised White Paper on *Fuel Policy* (Ministry of Power, 1967). Although the document was never formally debated – it was withdrawn for further consideration shortly after the devaluation of sterling – subsequent government statements and actions confirm that the broad lines of its policy have been accepted and, indeed, some of its specific proposals were embodied in the Coal Industry Act of 1967. Various forms of public assistance given to the mining industry and to mining communities faced with high levels of redundancy are but two examples.

The logic of the policy rests upon certain broad objectives which were first spelled out in 1965 – the need to maintain adequate and continuous supplies of fuel for the economy; the desire for a competitive level of energy prices compared with nearby western European economies; the preference for technically progressive and financially viable fuel industries; the goal of minimising foreign exchange expenditure; and the objective of ensuring freedom of consumer choice, provided the energy industries' costs are reflected in their prices – plus a set of energy demand forecasts. On this basis, assumptions can be made concerning the prospects for supplies and prices and a range of probable consumption outcomes can be derived. The economic and social effects of these probable outcomes can be assessed. Should the public interest require them, policies can then be formulated either to influence the pattern of fuel consumption or to ameliorate any social problems which seem likely to stem from the changing fuel demands. The strengths and imperfections of this approach towards fuel policy cannot be debated at this point.

Such an approach to a national energy policy naturally depends heavily upon a high quality of demand and supply forecasting, and in this connection a number of significant advances have been made in the understanding of the British energy market in recent years. The Ministry of Power, for example, has produced an integrated mathematical model of the fuel economy (Forster and Whitting, 1968) in the development of which the relationships between structural changes in the economy and energy demands, and between national economic performance and the behaviour of individual energy industries, have been more clearly exposed. Simultaneously, the Department of Applied Economics at the University of Cambridge (1968) developed a model of the demand component of the British energy market and continued with work on the supply side. But the point has to be recognised that the models of both the Ministry and Cambridge University are spatially aggregated descriptions of the national energy economy, simplifying it to the point where many of the geographical considerations to which attention has been drawn earlier in this chapter are generalised away. The suggestion cannot be avoided, therefore, that the demand and supply forecasts used as a basis for the 1967 White Paper are relatively insensitive both to the cumulative disadvantage of the coal industry, with its relatively wide range of spatial costs and its fragmented production–supply system, and to any major geographical changes in the energy economy. Such spatially aggregated models will inevitably bias demand forecasts in favour of those energy industries whose system costs vary most widely through space, and against those whose system costs vary the least. The faster that structural change occurs in those regional economies where the costs of a single and currently important source of energy are relatively high, the greater are likely to be the aggregate model's errors in forecasting. South Wales, Northern England and Scotland, it is relevant to reflect, are all Development Areas in which fairly rapid structural changes are likely to continue well into the nineteen-seventies (in large measure because of declining employment opportunities in the coal industry) and in which the delivered costs of coal are well above average.

TABLE 6.4. *United Kingdom: inland demand for energy 1966–9, and forecasts for 1970 and 1975 (million tons coal equivalent) (small discrepancies in the totals arise from rounding)*

	Actual				Forecasts		
	1966	1967	1968	1969	1970[a]	1970[b]	1975[a]
Coal							
Power stations	69.7	68.3	74.4	77.1	70	77	66
Gas works	17.2	14.8	10.9	7.0	3		—
Coke ovens	25.2	24.0	25.3	25.7	24		23
Industry	25.6	23.3	23.0	21.7	18	77	9
Domestic	26.9	24.5	23.6	21.9	21		15
Other inland	12.9	11.6	10.1	10.2	8		6
Total	177.5	166.4	167.1	163.7	144	154	120

TABLE 6.4 (*cont.*)

	Actual				Forecasts		
	1966	1967	1968	1969	1970[a]	1970[b]	1975[a]
Oil							
Power stations	12.6	12.8	11.1	14.3	13	11	20
Gas works	8.5	10.3	11.8	11.7	6		3
Road and air transport	30.5	33.1	35.4	36.8	38		47
Refinery fuel	7.9	8.1	8.7	10.8	10	116	11
Industry	36.6	38.6	40.7	43.6	41		43
Domestic	3.8	4.0	4.4	4.8	4		5
Other inland	13.6	14.3	15.9	26.1	17		18
Total	113.5	121.2	127.9	137.9	129	127	148
Gas							
Power stations	—	neg.	neg.	0.2	9	2	14
Industry	3.4	3.5	3.6	4.1	5		13
Domestic	7.8	8.8	10.3	11.7	9	15	18
Commercial and other	2.1	2.1	2.2	2.2	2		4
Total	13.3	14.4	16.1	18.2	25	17	50
Electricity							
Industry	39.4	35.4	36.4	38.6	41		52
Domestic and farms	36.6	37.0	38.6	41.7	47		57
Transport	1.3	1.4	1.4	1.5	2	107	1
Commercial and other	16.0	19.7	21.9	22.5	18		28
Total	93.3	93.5	98.4	104.3	108	107	138
Fuel use							
Coal and coke	70.3	68.9	74.8	77.3	69	77	66
Oil	12.6	12.8	11.1	14.3	13	11	20
Nuclear, hydro-electricity and net imports	10.4	11.8	12.5	12.7	16	16	38
Natural gas	—	—	—	0.1	9	2	14
Total	93.3	93.5	98.4	104.3	108	107	138
Total inland demand							
Coal	177.5	166.4	167.1	163.7	144	154	120
Oil	113.5	121.2	127.9	137.9	129	127	148
Nuclear and hydro-electricity	10.4	11.8	12.5	12.7	16	16	38
Natural gas	1.1	1.9	4.4	8.5	25	17	50
Total	302.5	301.4	311.9	322.8	315	315	356

(a) April 1967 estimates of Ministry of Power.
(b) November 1967 estimates of Ministry of Power.
Sources: Ministry of Technology, 1970; Ministry of Power, 1967.

The April and November 1967 estimates by the Ministry of Power for energy consumption in 1970 and 1975 are reproduced in Table 6.4, along with actual data for the years 1966–9. The April 1967 forecasts were based upon an assumed 3 per cent per year increase in the gross domestic product, upon an expectation that the relativity between coal and oil prices would remain unchanged, and upon the proposition that 10 per cent of the natural gas from the North Sea would be used in power stations. The level of coal demand was based upon the assumption that the existing government measures to help the industry would be retained until the early nineteen-seventies: these included the ban upon coal imports, the £1.96 per ton tax upon oil, and the direction of the CEGB to burn coal rather than oil in certain power stations. The size of the nuclear power programme was conditioned by the government's 1967 estimates of its comparative and prospective costs and upon a firm commitment to the advanced gas-cooled reactor construction programme. The natural-gas figure was based upon the best estimate of availabilities in 1970 and 1975. Oil was treated to a considerable extent as a residual and flexible source of energy which would serve the rest of the forecast requirement for 315 million tons of coal equivalent energy in 1970 and 356 million tons in 1975.

However, the prospect of a continuing replacement of coal by alternative energy sources so as to reduce its inland market to 142 million tons in 1970 was judged to be undesirable. The NCB took the view that a contraction of 30 million tons between 1966 and 1970 would cause unmanageable difficulties in the industry and convinced the government that additional short-term measures were required to protect coal. It was agreed that the industry should be supported at the 157 million ton level in 1970. On the assumption of 3 million tons of exports, this implied increasing the market for coal by 10 million tons. The only practicable way in which this could be achieved was by asking the CEGB, and with reference to much smaller tonnages the gas industry, to burn more coal. This policy, therefore, served as the basis for the November 1967 estimates of the energy market in 1970, which also (and without explanation) revised downwards the prospective availability of natural gas. These November 1967 estimates formed the basis of the policies proposed in the second White Paper on *Fuel Policy*.

It will be noted that the total demand for energy in Britain has exceeded expectations between 1966 and 1969. The demand for energy in 1969 was in fact nearly 6 per cent higher than the forecast level for the following year. Within this slightly larger market, natural gas consumption appears to be reasonably well in line with the November 1967 forecast; although

only 8.5 million coal equivalent tons were consumed in 1969, it is widely
expected that over 16 million coal equivalent tons will be used in 1970.
Nuclear power, on the other hand, has clearly fallen behind the expectations of 1967. Partly this is a consequence of a series of technical faults in
the Magnox reactors which has reduced their output; but more particularly it is the result of serious delays in the commissioning of the Wylfa Head
and the Dungeness 'B' reactors. The coal industry, therefore, has been
presented with an unexpected opportunity to benefit from the larger
aggregate market for energy in Britain than was expected in 1967, from the
relatively slow arrival of natural gas and from a temporary set-back in the
nuclear power programme. Coal consumption in 1969 was some 164 million tons, well up to the forecast level. However, about 8 million tons of
this total were withdrawn from stocks and production for inland demand
was less than 152 million tons. In the financial year 1969/70, production for
inland markets had fallen yet again to 148 million tons, and for the calendar
year 1970 it could well be less than the April 1967 estimate of 1970 demand,
more especially following the withdrawal of labour on several coalfields in
November of that year. Even with further withdrawals of coal from the
industry's shrinking stocks, consumption in 1970 is unlikely to exceed
155 million tons. The irony of the coal industry's situation at the beginning
of the nineteen-seventies, therefore, is that the steady contraction of the
industry over the previous decade, and in particular the run-down of its
labour force, plus its fragmented production–supply system, have denied
it the ability to meet all of its potential demands even in a protected market,
let alone respond vigorously to unexpected market opportunities.

The real beneficiary of the situation, therefore, has been the oil industry.
Already in the first quarter of 1968, oil consumption had expended to an
annual level slightly in excess of the November 1967 demand forecast for
1970. Towards the end of the latter year, not only was oil consumption
for the first time running at an annual level in excess of 150 million coal
equivalent tons, but it was also for the first time above that of coal, and had
passed the 1967 forecast for consumption in 1975. There are no grounds
for believing that this expansion of the oil industry's markets will not continue. By implication, therefore – and in particular assuming that the
nuclear power programme returns to its earlier schedule, and the gas industry
is able to dispose of its committed supplies of natural gas for the middle
nineteen-seventies – it appears highly unlikely that the coal industry will
have an inland market for 120 million tons in 1975. Even 100 million tons
now appears to be somewhat optimistic unless there is a continuing high
level of protection.

Such a conclusion is endorsed by a consideration of recent developments in the electricity generating industry, an industry which in the past more than any other primary fuel user has been subject to powerful political pressures to burn coal. Despite the fact that coal – as early as 1962 in Scotland, and from about 1964–5 in England and Wales – has been marginally more expensive than oil at a large number of power stations, some 75 per cent of the country's electricity was still generated from coal in 1969. Oil was the primary fuel for about 15 per cent. By that year, however, it was clear that the managements of the electricity generating authorities were determined to reduce as quickly as possible their dependence upon coal – for which the GEGB was paying an average price of over 2.1 pence per therm. Doubts concerning the adequacy of coal supplies over the winters of 1970–1 and 1971–2 undoubtedly shifted the government's stance. Oil provided an immediate alternative. At mid-1969, there were 15 power stations in the United Kingdom burning heavy fuel oil, and a further four large ones were under construction at Fawley, Milford Haven, Kingsnorth on the lower Thames and Inverkipp on the lower Clyde (Figure 6.4). Since then, permission has been requested and granted for the conversion of nine existing coal-fired power stations to oil – five of these are in London or on lower Thamesside, two are near Glasgow and the other two are near Bristol and Cardiff – and for the construction of a new 3,300 MW oil-fired station at the Isle of Grain. In addition, permission has been requested to convert four more existing stations to oil – in Portsmouth, Burnley, Leeds and Edinburgh – and to build new oil-fired stations at Plymouth, Ince, Killingholme, Littlebrook (Kent) and Brunswick Wharf on the site of the closed East India docks. Further, the CEGB has been granted permission to convert Hams Hall 'C' power station to dual coal–natural gas firing, and has contracted to purchase 300 million therms per year from the Gas Council on an interruptible basis from 1971; it has applied to make a similar conversion at the West Thurrock plant near Tilbury. Plans for the construction of a further coal-fired power station in the East Midlands at West Burton have been abandoned.

The new oil-fired power stations will claim a high (base-load) place in the merit order of the Generating Board's production facilities; and the converted stations, rising in the merit order as a result of their improved operating costs, will displace a larger tonnage of coal from the electricity industry's primary fuel market than was consumed by the stations before conversion. From a consumption of just over 14 million coal equivalent tons of oil in 1969, it now appears likely that at least 22 million coal equivalent tons will be burned in British power stations in 1971; the approval by the

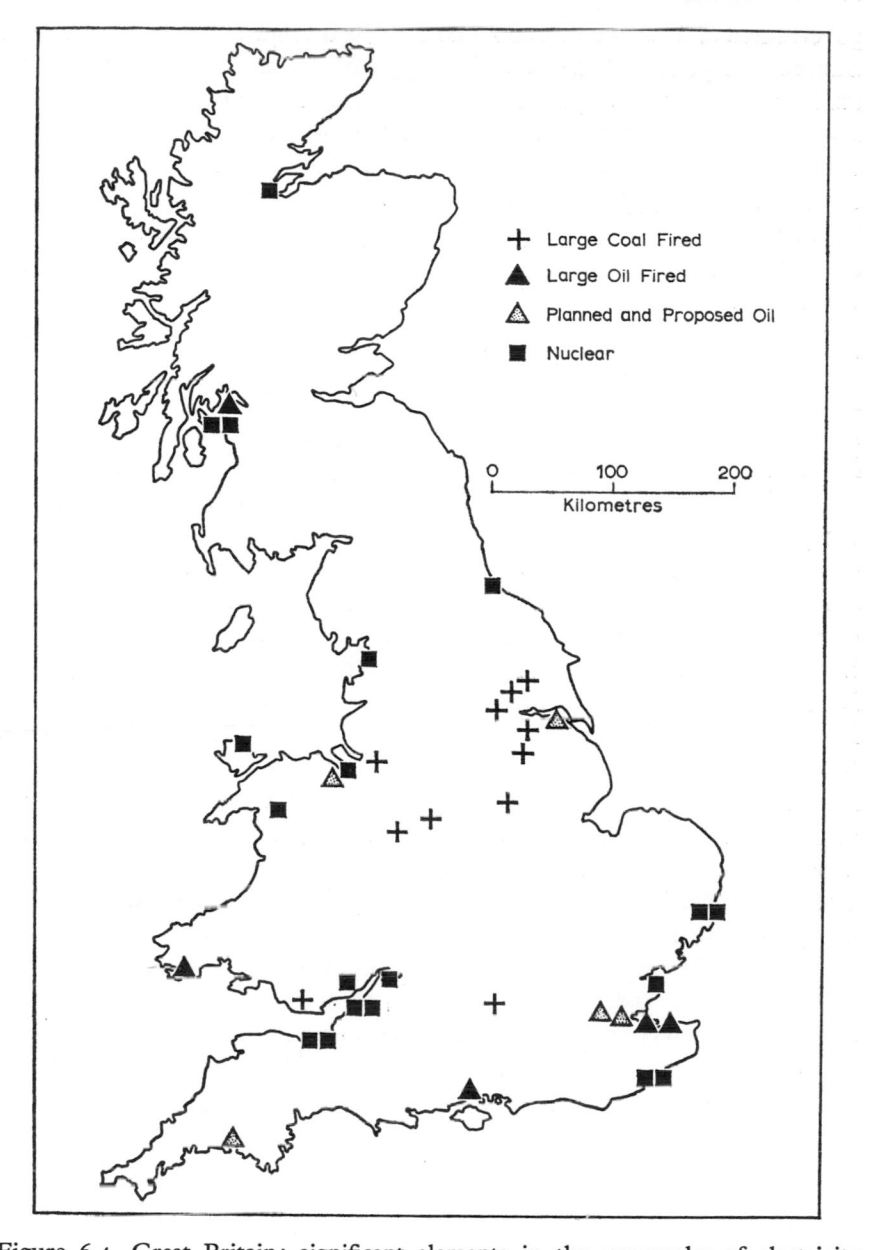

Figure 6.4. Great Britain: significant elements in the geography of electricity generation, *circa* 1970.

Sources: Central Electricity Generating Board, South of Scotland Electricity Board, North of Scotland Hydro-Electricity Board.

government of further conversions could raise the figure even higher in 1971. Subsequent years will see the Fawley, Milford and Kingsnorth power stations commissioned plus a further increase in the quantity of oil consumed by the industry. The removal of the coal-burn subsidy in 1971 can only further endorse the trend. Thus, the April 1967 Ministry of Power forecast that in 1975 only 20 million coal equivalent tons of oil would be used for generating electricity is undeniably a gross underestimate. By the same token it is very difficult to believe that the generating authorities will provide a market for 66 million tons of coal in that year.

There are of course various reasons why the 1967 forecasts of the Ministry of Power for coal production and consumption in the middle nineteen-seventies now appear to be more than a little over-optimistic. Not least is the much greater effect which the rapid wage inflation of 1969–71 has had upon the costs of the coal industry by comparison with other energy industries, as a result of the former's much greater use of labour at every stage of its production–supply system. To recognise the importance of such factors, however, is not to deny the probability that, had the spatial characteristics of the British energy market been more carefully examined in 1967, a more realistic set of demand and supply forecasts could have been produced for the several energy industries. Spatial competition is becoming more, not less, acute. Therefore, for forecasts made now and in the future it is increasingly necessary to incorporate the spatial element. With the data currently available, and in the present state of the economic model building and the associated forecasting art, the addition of a spatial dimension to the existing crude models of the national energy market will be exceedingly difficult. Whilst they are denied even approximate mathematical expression – as are other equally important considerations in many forecasting exercises – for some time intuitive spatial qualifications to the steadily improving aggregated forecasts of the energy economy will have to be applied and will have to suffice. But within the next decade, and with the improvement of geographical data and understanding, the spatial dimension must be incorporated within the models of the energy economy in order to improve the accuracy of their description and the insights of the forecasts upon which public policy must rest.

References

Adelman, M. A. (1964a). 'Oil prices in the long run (1963–75)', *Journal of Business of the University of Chicago*, XXXVII, 143–61.

Adelman, M. A. (1964b). 'The world oil outlook', in M. Clawson (ed.), *Natural Resources and International Development*, Johns Hopkins.

Berrie, T. W. (1967). 'The economics of system planning in bulk electricity supply', in R. Turvey (ed.), *Public Enterprise*, Penguin.

British Railways Board (1963). *The Reshaping of British Railways*, HMSO.

Brunner, C. T. (1962). 'Productivity in oil distribution in Britain', *Institute of Petroleum Review*, January, 5–10.

Department of Applied Economics, University of Cambridge (1968). *The Demand for Fuel 1948–1975*, Chapman and Hall.

Forster, C. I. K. and I. J. Whitting (1968). 'An integrated mathematical model of the fuel economy', *Statistical News*, November, 1–6.

Hauser, D. P. (1969). *Postwar Power Station Location and Interregional Fuel and Electricity Flows in England and Wales*, unpublished PhD thesis, University of Cambridge.

Hubbard, M. E. (1963). 'Productivity in the petroleum industry – the sources of some of its improvements', *Journal of the Institute of Petroleum*, LXLIX, 29–57.

Manners, G. (1959). 'Recent changes in the British gas industry', *Transactions and Papers, Institute of British Geographers*, XXVI, 153–68.

Manners, G. (1964). *The Geography of Energy*, Hutchinson.

Manners, G. (1965). 'The changing location of secondary energy production in Britain', *Land Economics*, XLI, 317–24.

Ministry of Power (1965). *Fuel Policy* (Cmnd. 2798), HMSO.

Ministry of Power (1967). *Fuel Policy* (Cmnd. 3438), HMSO.

Ministry of Technology (1967). *Nationalised Industries – A Review of Economic and Financial Objectives* (Cmnd. 3437), HMSO.

Ministry of Technology (1970). *Digest of Energy Statistics 1970*, HMSO.

National Board for Prices and Incomes (1969). *Gas Prices* (Cmnd. 3924), HMSO.

Odell, P. R. (1966). 'A three point approach necessary to exploit sea gas', *The Times*, 21 July.

Odell, P. R. (1968). 'The British gas industry: a review', *The Geographical Journal*, CXXXIV, 1, 81–6.

Pratten, C. and R. M. Dean (1965). *The Economies of Large-Scale Production in British Industry*, Cambridge University Press, Department of Applied Economics, Occasional Papers No. 3.

Rawstron, E. M. (1955). 'The salient geographical features of electricity production in Great Britain', *Advancement of Science*, XII, 75–82.

Robens, Lord (1970, 1969, 1968). Addresses to the annual conference of the National Union of Mineworkers, mimeographed by the National Coal Board.

Schurr, S. H. (1960). 'Foreign trade policies affecting mineral fuels in the United States and Western Europe', paper presented to the American Association for the Advancement of Science, New York.

Shepherd, W. G. (1964). 'Cross-subsidizing and allocation in public firms', *Oxford Economic Papers*, XVI, 132–60.

7. Growth, technical change and planning problems in heavy industry with special reference to the chemical industry

KENNETH WARREN

The terms 'heavy industry' or 'light industry' are normally used far too loosely. Definition proves much harder than the quick wave of the hand over a map or the glib labelling of a district would suggest. Transistor manufacture and the production of stationery are clearly enough on one side of the fence; shipbuilding and the production of large pressure vessels on the other. Size, volume and weight of product are important in the classification for they affect the type of input, the material handling equipment of the plant and the minimum size of the whole operation. When large masses of material, perhaps hot or corrosive, are involved much investment in plant is required; characteristically, therefore, heavy industry has a high capital/labour ratio, and, in spite of big labour forces, a high value of output per employee. In general products are sold not to individuals but to other manufacturing industries. Such indices of heaviness are admittedly crude, for many industries or individual firms span a much wider range than can be comprehended under the title of 'heavy' – the big chemical firms for instance make pharmaceuticals and gardening requisites and the steel industry's products range all the way from giant castings and forgings to the razor blade steel and umbrella frames of one major Sheffield plant. A number of heavy industries, some new, some long established, can be readily isolated, bearing some, but not in each case all, of these characteristics. The major ones are steel and engineering, including not only shipbuilding but also the motor industry (though the latter is out of line in some respects), machine tools, oil refining and the production of heavy chemicals.

These trades share certain requirements, of site, of location and for labour, raw materials and markets. They make a very marked impression on their neighbourhood for their operations require much space and involve much movement, accompanied frequently by noise and pollution. As such industries have very large outlays for a big site, buildings, massive equipment and the establishment of a labour force with specialised skills, they commonly are victims of inertia – of commitment to a location once it has been developed – even when marketing situations change and technical advance upsets old processes and alters the patterns of supply. Their large work forces, their role as key elements in industrial complexes and their frequent concentration in old, problem industrial areas rather than on metropolitan industrial estates are characteristics which make them of great significance in regional economic planning.

In Britain, some of these heavy manufacturing industries are old and their location pattern has ossified. Although the techniques of shipbuilding and therefore its calls on the environment have completely changed, no major new building location for ships has been developed since World War I. Instead, old yards have been modernised even though this is very much a second best and may ultimately prove a dead end (*Economist*, 1968*a*). Greenfield sites were frequently used by the expanding steel industry of the USSR and still are in the case of Japan, but they have been very rare in twentieth-century Britain. Corby, internationally acclaimed in the nineteen-thirties, continues to expand, though its location is now a liability. In the case of new or rapidly expanding heavy industries, however, changing conditions can be reflected in the choice of new types of location. This is clearly the case in oil refining, where capacity went up from 2.4 million metric tons in 1938 to 82.5 million in 1968. Even in this case, the strength of inertia proved substantial. Six of the fifteen refineries of over 1 million tons capacity at the end of 1968 existed as refineries or as major oil depots in the nineteen-thirties, and these six had well over half the national throughput capacity (Petroleum Information Bureau, 1969).

The chemical industry illustrates particularly well many of the characteristics and problems of heavy industry. It has revolutionary technologies displacing traditional ones, rapid growth, keen competition and both old and very new locations. Its impact on the environment is impressive and frequently troublesome. In such an industry, the importance of making the right choice of both location and site is great. However, what is economically right for the firm may not be best for the nation, the region or the locality. Sometimes there is clearly scope for a more positive Development Area policy but the form of intervention remains in dispute.

TABLE 7.1. *United Kingdom: chemicals as a proportion of all manufacturing,*
1963 and 1968 (chemicals as a percentage of the total)

	1963	1968
Net output	9.3	10.1
Export	11.5	12.4
Net foreign trade balance	10.2	14.7
Investment	13.0	15.0
Numbers employed	5.3	5.1

Source: Chemicals EDC, 1970.

Characteristics of the chemical industry

Chemicals are a pronounced growth sector within the heavy industrial economy. It is reckoned, for instance, that consumption of petroleum-based organic chemicals in the non-communist world increased from about 1.5 million tons in 1948 to some 50 million in 1970, and is likely to grow by 10 per cent annually to reach 250 million tons by 1985. Western European ethylene production was 70,000 tons in 1950 but 5 million tons in 1969 (Shell, 1970). In some countries, other heavy industries are in the growth industry sector but in Britain chemicals are outstanding. Between 1960 and 1966, the annual compound growth rate for British chemical output was 5.3 per cent as compared with 2.8 per cent for all manufactures and 2.7 per cent for the Gross Domestic Product. In the early nineteen-fifties, it was the fifth British industry as measured by sales volume; by 1962, third and by the late nineteen-sixties second. At this time, it had 10 per cent of the turnover of British manufacturing industry (Table 7.1): by 1980, it is estimated as likely to have 15 per cent (Economic Research Group, 1967; Iliff, 1969). By all indices chemical manufacture is a heavy trade. Its major firms, individual works and units of plant are all big (Tables 7.2, 7.3, 7.4). In 1963, only 3.5 per cent of all British chemical establishments employed more than 1,000 workers but these had 39.1 per cent of the total labour force. Capital stock per worker is especially high: in 1965, the average in United Kingdom manufacturing and construction was £2,260 but in chemicals £6,850 and annual turnover per worker was £5,570 (Economic Research Group, 1967, 8, 9, 15). In petrochemicals the investment is sometimes in excess of £25,000 per worker. Overall, at the end of the nineteen-sixties, when British chemicals produced 10 per cent of the output of manufacturing and construction trades, it had only 4 per

TABLE 7.2. *United Kingdom: industrial structure of general chemicals and all manufactures, 1963*

Industry and establishment size group	Number of establishments	Net output (£ million)	Total employment (thousands)
General chemicals			
Under 1,000 employees	811	186	78
Over 1,000 employees	26	200	60
(of which over 10,000)	(7)	(107)	(28)
Total	837	386	138
All manufacturing			
Under 1,000 employees	83,037	6,434	5,044
Over 1,000 employees	1,189	4,275	2,795
(of which over 10,000)	(23)	(480)	(324)
Total	84,226	10,709	7,839

Source: *Census of Production*, 1963, *General Tables* (Board of Trade, 1969).

TABLE 7.3. *United Kingdom: concentration of production in large enterprises, chemicals and other heavy trades, 1958 and 1963 (percentage of total industry sales by five enterprises with largest sales)*

	1958	1963
Chemicals		
Inorganic	62.8	63.1
Organic	59.3	64.9
Fertilisers	—	78.6
Pig iron	54.7	67.3
Steel ingots	84.2	76.6
Textile machinery	64.7	60.3
Switchgear and switchboards	59.7	59.0
Private cars	90.4	91.5
Commercial vehicles	76.1	84.1

Source: *Census of Production*, 1963, *General Tables* (Board of Trade, 1969).

TABLE 7.4. *United Kingdom: net output, wage payments and transport costs for three basic industries and all manufactures, 1963*

	Net output (£ million)	Wages for operatives		Transport costs	
		(£ million)	(% of net output)	(£ million)	(% of net output)
All manufactures	10,851	3,990	36.7	568	5.2
General chemicals (1)	387	70	18.0	28	7.2
Iron and steel (2)	405	176	43.4	31	7.6
Engineering and Electrical goods (3)	2,507	906	36.1	61	2.4

(1) Standard Industrial Classification 271 (3).
(2) Standard Industrial Classification 311.
(3) Industrial Order VI.
Source: *Census of Production*, 1963, *General Tables* (Board of Trade, 1969).

cent of the work force but almost 15 per cent of the total investment (Iliff, 1969; OECD, 1970). Current rates of investment are high, some £200–300 million a year, and at their Wilton works on Teesside Imperial Chemical Industries (ICI) have at peaks of development been investing £1 million a week. In the mid-nineteen-sixties, two-thirds of the chemical output was sold to other industrial concerns. These global characteristics however hide a terribly complex situation.

Some chemical products are made directly for the consumer and in small plants. Drugs and toilet preparations are a world away from naphtha crackers and ammonia columns. By 1969, ICI had an average labourforce of 1,800 in each of its 80 plants but some 30,000, or over one-fifth of its labourforce, worked in the two Teesside complexes of Billingham and Wilton in whose development some £500 million has been invested. Some product lines have been growing much more quickly than others – plastics and organics (those with a carbon base) much more than inorganics, though ammonia is an exception, not only as a fast growing inorganic but, and confusingly, as an inorganic now made from a hydrocarbon feedstock. New materials and products have displaced old ones, techniques are continually being modified and occasionally more fundamentally altered, and the scale of plant has been increasing in many fields with amazing rapidity.

Finally, the question of demarcating a 'chemical industry' is extremely difficult. Three-quarters of a century ago, chemicals were already produced by coke ovens, tar distilleries, nonferrous smelters and iron and steel-works. The British Steel Corporation has inherited some of these important chemical operations. In the inter-war years, textile firms entered the trade with the manufacture of materials for artificial fibres and output has widened from rayon to the legion of their present productions. Since 1945, the major oil companies have built petrochemical plants in association with their refineries. It is true that for Shell or British Petroleum, chemicals still represent only about 12 per cent of turnover but the proportion is growing and their products are already competing directly with those of chemical companies such as ICI or Fisons.

The industry whose structure is so complicated and whose links with other industries are so many nonetheless has distinctive characteristics. These were summed up neatly by the Economic Research Group (1967, 4):

Whereas other branches of industry generally modify only the shape of primary materials, the chemical industry effects a true change of substance, and often it is possible to obtain the same product both from different raw materials and by different processes...The materials employed...are normally in liquid, gaseous or powder form and processing takes place by flow, mixing, compression or similar action, often involving the use of considerable energy. A typical chemical plant usually appears from the outside to consist of tanks, silos and piping, the capacity of which in terms of flow and volume can be varied with ease. In view of these production conditions the scope for automation, expanding capacities and cost reduction is frequently greater in the chemical industry than in other production fields...Most chemical processes produce joint products.

Traditional materials and plant distribution

The old heavy chemical industry was based partly on home, partly on imported materials (Hardie and Pratt, 1966). Sulphuric acid was and is made from imported sulphur and pyrites though over the last forty years important tonnages have been derived from the anhydrite ($CaSO_4$) of Teesside and, more recently, of West Cumberland. Phosphates, for phosphorus manufacture and fertilisers, are imported largely from North Africa, and potash has come from the continent, though within the next few years very important domestic sources, now being opened in Cleveland, will transform Britain into an exporter. Nitrogen, formerly obtained from Chilean deposits of sodium nitrate has long been derived by fixation from the atmosphere.

The three chief domestic chemical raw materials have been coal, salt and limestone. Coke ovens have supplied the coke which has been vitally important as a reducing agent in the production of water and producer gases and for the manufacture of calcium carbide. The carbonisation process also produced coal tar from which, as the gas industry's advertising used to describe, a host of chemical treasures were derived. These included the aromatic chemicals the chief of which, benzene, was in turn a source of dyes and pharmaceuticals. In spite of the rapid advance of oil-based aromatics, Western Europe ended the nineteen-sixties with some 0.6 million tons of coal-based benzene capacity (*Economist*, 1968b). Ammonia compounds and, most important of all, sulphate of ammonia were also derived from the coke ovens. The salt fields, chiefly that of the Mersey Basin, supported major inorganic chemical manufacture and especially the production of soda ash and chlorine. Limestone from Pennine quarries was brought by rail and later also by road to the lowland chemical districts without much trouble at a time before amenity problems began to loom and before the right of the individual to walk the wilderness undisturbed had been established.

From the heyday of the Leblanc alkali trade in the third quarter of the nineteenth century, the Mersey Basin held the biggest concentration of heavy chemicals in the United Kingdom. There were major nodes of activity at St Helens, Runcorn, Widnes and, after the introduction of the Solvay process, at Northwich as well. Meanwhile, the Tyneside alkali trade, equal in status in the early eighteen-sixties, declined and died out completely; but in the nineteen-twenties the foundations were laid for an even bigger focus of chemical manufacture in the North East of England, this time on the Tees Estuary. In dyestuffs, explosives, paints, soaps, fertilisers and drug manufacture, many other centres of chemical manufacture were being established throughout Britain. They have no physical common denominator like the ingot ton of the steel industry or the gross tonnage or steel throughput of the shipyards. Their only shared unit is value of output and the difficulty of finding this on any spatial basis more disaggregated than census of production regions presumably explains the absence of early general maps of chemical production in Britain. Each of the main industrial areas had some share in the heavy chemical trade but by World War II the main foci were on the Mersey and the Tees. The South East region was the leading chemical producer, both in terms of employment and net output, at the time of the last Census of Production. Its pre-eminence was however largely due to the production of fine chemicals, though it has some heavy lines too. In the mid-thirties, about 10 per cent of British heavy acid production came from the London County Council area or along the Thames up to 19 kilometres (12 miles) east of Charing Cross. In the late nineteen-forties, Fisons considered Thamesside for their biggest post-war development. Largely because of fears of problems associated with the proximity of built-up areas and possible complaints about damage to amenity, they decided in favour of Immingham (Luttrell, 1962, 654–7). This long-established dominance of the development areas in heavy chemicals is of great significance to the present situation of the industry.

Where know-how, a labourforce and existing equipment are available, new small-scale processes can be most easily attached to existing manufacturing units. When full-scale production is decided upon, it is common to fit the production into existing works to share common services – internal economies, or 'offsites' as they are called in the terminology of the plant engineer – and so to take its place in the product flows of a chemical complex. Wholly new plants are therefore rare and usually built at a relatively early stage of economic growth. In recent decades, this marked tendency to perpetuate old patterns has been reinforced by a wish to minimise the effect of redundancy in old-style operations. However, new entrants to the trade – a class, which in a capital-intensive business such as heavy chemicals, is inevitably small – and firms moving in from major interests in a related business will be free to choose new locations. For example, the entry of the oil companies to chemical manufacture contributed a new impetus to growth and fresh elements to the location pattern.

As late as 1938, the oil refining capacity of the United Kingdom was only 2.4 million tons; by 1970, it was in excess of 117 million tons. The early post-war expansion was oriented to the major centres of national demand, as at Shell Haven or Fawley, but later developments have shown a greater range of locational influence. Examples are the search for deepwater at Milford Haven, or the Firth of Clyde, new market locations as at Canvey Island or compromise locations such as those on Humberside and at Teesport. In some cases, as with recent Milford Haven developments or the 'chemical' refinery proposed for Invergordon – and again deferred in 1970 – international marketing is a consideration as, or even more, important than the national pattern of demand. By 1968, seven of the fifteen oil refineries of over 1 million tons capacity had associated petrochemical operations.

The first oil-based chemicals were made in New Jersey in 1919 but it was another thirty years before production in the whole of Western Europe topped 100,000 tons a year. In 1950, about 9 per cent of British organic chemicals were made from oil but by the late nineteen-sixties as much as 70 per cent. Some of this growth has been based on purchased feed-stock, as in the greatest complex of all, Wilton–Billingham, where there was no local refinery until 1966. In most cases, however, development of petroleum-based chemicals has been dependent on entry by the oil concerns into chemicals manufacture. Shell began in a small way at Stanlow in 1942 and later at Partington/Carrington near Manchester and at Shell Haven. British Petroleum and the Distillers Company, as British Hydrocarbon Chemicals, entered the trade at Grangemouth (1947) and at Balgan Bay near Port Talbot in 1963. Esso's Fawley complex first made chemicals on a large scale in 1958. By 1969, Shell estimated that petroleum feedstock provided 85 per cent of the organic base chemicals and 81 per cent of the ammonia produced in the non-communist world.

Economies of scale and of integration in chemical production

Chemical plants have long been big operations; even more noteworthy has been the rapidity of the increase in scale. These characteristics cause locational inertia and make the choice of location for the few wholly new plants of great economic and planning significance. Occasionally, a major technical change is the chief cause of a sharp upward movement in plant size, as with the new chloride process for titanium oxide which requires a much bigger initial plant than the older sulphate process; more commonly, increase in plant size is associated with progress in process plant engineering and the recognition that capital outlay, energy consumption, space requirements and labour input do not increase in proportion. The most impressive illustrations have come from the production of ammonia and of ethylene, the chief olefine and basic material for plastics, synthetic fibres and rubber. Until the early nineteen-fifties most ethylene made in Britain was based upon molasses, but since then use of naphtha has become common. Naphtha crackers have increased in size remarkably quickly – by a factor of 10 to 15 in the last fifteen years (Iliff, 1969). British Hydrocarbon had three ethylene units at Grangemouth in 1965 – two with a capacity of 30,000 tons a year and one of 60,000 tons. It was then announced that these would be put in reserve to be replaced by one unit of 250,000 tons. Early in 1969, ICI completed a 450,000 tons ethylene plant at Wilton. When this was building in 1967 it was expected that its 82 m × 5 m column would give ten times the output of a 46 m × 1.5 m column at unit costs of only 40 per cent. If the operation could be started up without a hitch, the return on the investment was expected to be twice that obtainable on three units of 150,000 tons (ECN, 1965*a*, 1969*a*; *Economist*, 1967; 1968*c*; Booth, 1969).

Production of ammonia has a much longer history but in Britain is similarly now largely based on naphtha. By 1967, Billingham production costs were already reckoned no more than one quarter the costs when coal was the raw material a decade before (ECN, 1967*a*). ICI put the capital cost of the three 300,000 tons ammonia units built there in the late nineteen sixties at only £25 per ton of annual product as compared with £35 in the 100,000 tons plants which they replaced. Kellogg International, the plant engineers for the new plant, indicated that whereas the largest single train ammonia installations built to 1960, having an output of 300 tons per day, had production costs which at best were about £14 a ton, the new units, of capacity from 1,000–1,500 tons, could produce for half this cost (ECN, 1967*b*). The new Billingham installations each had a unit manpower requirement less than one-thirtieth that of the plant they replaced and unit power consumption was under one-sixth (Chambers, 1968). As international competition becomes keener the incentive to install bigger plant increases. In 1967, the industry's 'Little Neddy' concluded that productivity in the United States chemical industry was 2.5 times the British level and that two-thirds of the difference could be attributed to scale economies. The rest of the difference was the reward for better division of labour, the availability of more specialist assistance for managers and foremen and a greater profit consciousness throughout (Chemicals EDC, 1967).

The implications of the search for scale economies are very wide ranging. For the individual firm there is stress on growth to accommodate the new, highly productive units, and also to make room for progress in processing technology. Even before its 200,000 tons a year ethylene cracker was completed, ICI was planning the 450,000 tons unit (*Economist*, 1966). This leads to the danger of over-capacity and to a keenness in competition which can only be met by installing similar high-output/low-cost plants. Standing charges on these are very high, not only because of the very large outlay but also because of their rapid obsolescence; ten-year depreciation schedules are now general, with five to seven years being more usual in a few cases. Fortunately for the industry, amendment to the Restrictive Trades Practices Act has permitted discussions designed to avoid the dangers of gross over-capacity and Shell took full advantage of this relaxation in discussions with ICI and British Petroleum before announcing the £225 million oil and chemical expansion for north-western England early in 1970. However, as marginal costs – the cost of merely covering operating outlays – are low, sometimes less than half the total costs, it is possible to operate the plant at a high level by selling overseas at well below home trade prices. Thus, the 0.9 million tons Billingham ammonia plant has a capacity well in excess of the needs of the British fertiliser industry, its chief outlet, and in the supply of which other important ammonia producers compete. As ICI's chairman noted, the Billingham installation could only be justified on the assumption that much of its output would be exported. Similarly, marginal costs in their £15 million ethylene unit are low enough to justify the export of some of its output for the first four or five years (Chambers, 1968; *Economist*, 1969).

A corollary of this is the recognition that the British market is already too small to support an optimum-sized chemical plant in some product lines. It is sobering to note the remark of the first president of ICI America, '– the U.S. chemical industry grows by more than one I.C.I. every year' (*Chemistry and Industry*, 1969 a). Movement into the European Economic Community is seen as essential to the continued growth of the bigger concerns. British production of chemicals in 1965 was valued at £2.5 billion but that of the Community at £7.8 billion. For many years the ICI export business with Europe has been increasing by 15 per cent annually (*Economist*, 1966 b; Bell, 1967; Chemicals EDC, 1970, 3). As well as causing keener competition, the growth of giant units has been accompanied by growing exchanges of materials between plants. ICI has not yet built any of the most basic – the 'building block' – chemical capacity in Europe. Instead, its major works at Oestringen (Heidelberg) is supplied from Rozenburg (Europoort) which in turn receives ethylene, its main raw material, from Teesside. However, from 1970 ICI is connecting the Rozenburg works to the ethylene grid which links four other major producers or consumers at Pernis, Terneuzen, Antwerp and Jemeppes. Through this pipeline, major exchanges will be possible so that each concern can avoid making major extensions of capacity merely to meet its own needs (ECN, 1969 b, 1970 a). Another long-established interchange involves shipment of olefines from Shell at Pernis to Billingham, where it is processed into plasticiser alcohol, much of which is shipped back to Holland for European sale (*Chemistry and Industry*, 1967). Similar exchanges have been developed in Britain and are examined below in connection with the concept of chemical complexes.

The new, extensive semi-product exchanges – along with growth in British foreign trade in chemicals, European Economic Community economic growth and possible United Kingdom membership of the Community – point to a long-term dilemma in the location of new capacity. If British chemical companies eventually find it more attractive to concentrate their new investment in Europe, even to supply British demand, what attitude should the government adopt? On the other hand, today's immense scale economies combined with bulk water or pipeline shipments make it possible to supply international markets from an existing plant that is well located. The very large investment assistance to industry in the Development Areas has encouraged this polarisation. However, as competition becomes keener and British interest in continental chemical production increases, it may become desirable for big companies to build basic capacity as well as finishing plants there. A sharp reduction in Development Area investment grants would presumably hasten the time at which this would become an attractive policy.

Diseconomies of scale lie in wait to trap the unwary or unfortunate seeker after economies and so act to check the concentration of production in super units. The long period that elapses from the initial ideas about the construction of a major unit to its commissioning is one cause of the difficulty. It has been suggested that planning a major installation such as a naphtha cracker or ammonia plant may take at least twelve months. This stage of commercial assessment involves a choice of process, the consideration of complementary products, an exploration of engineering problems and the presentation of the case to the Board. An additional step recently added to this sequence for really big projects involves consultation with other producers. Plant process engineers require at least thirty months from the Board's approval to the start of production. Commissioning will take six months or so – a total from the beginning of planning of three and a half to four years. It may take far longer. Shell's £225 million development in the north west was reported to have been under consideration for three years before it was announced in January 1970 and the whole project will not be on stream until 1975 (Fenning, 1968; *The Times*, 1970a, b). As costs increase, it is vital that this period should be minimised or the whole project is imperiled. Back in 1965, ICI had £200 million or a fifth of its total capital invested in plants which were either incomplete or not yet in full operation and therefore adding to costs rather than contributing to profits (Chambers, 1966). Between October 1966 and May 1970 the costs of British Petroleum's Baglan Bay expansion rose by some 18 per cent, largely due to wage inflation in the construction business. The significance of delay in completion may be assessed by taking into account the estimate that the discounted cash flow return on a big project may be reduced by as much as one to two per cent for every six months lag behind schedule (*The Times*, 1970c; Fenning, 1968). In fact delay in plant completion is common. There are various reasons. One of the more important is the difficulty in adjusting supply of plant engineering facilities to meet the sudden surges in demand for their services.

Demand for chemicals grows quickly though fairly smoothly but today's big units of plant make major expansion of capacity and calls on the process engineers come in steps. British Petroleum at Grangemouth and ICI at Wilton each brought a new ethylene plant into production in 1969; the result was an increase in national capacity from 925,000 to 1,650,000 tons (*Economist*, 1968 c). The chemical firms compete for the plant engineers' services with the oil refiners and for both trades there are major booms and recessions in new construction – ICI spent £145 million on capital account in 1966, £81 million in 1967 and £90 million in 1968 (*Economist*, 1968 d; ICI, 1968). In periods of advance, delivery dates lengthen considerably. Shell has expressed a wish to rationalise the work load for the total of no more than four or five contractors in the world which can build ethylene crackers of the size involved in its north-western expansion. Shell has good reason for caution on this point of completion. In 1965, its joint fertiliser concern with Armour, known as Shellstar, decided to build a 0.75 million ton plant at Ince Marshes near Ellesmere Port. By the time it was in production in 1970 it was two years late, largely as a result of labour troubles on the site, and costs had increased by £6 million, or more than a third. The effect of this on the discounted cash flow return of the Ince Marsh operation is unknown but obviously highly unfavourable.

After their commissioning, another problem with big plant units concerns the time involved in running through the almost inevitable teething troubles and into the period of high operating rates. Wilton's 450,000-ton ethylene plant had a smooth start-up in 1969 but the three 300,000-ton Billingham ammonia units, costing £30 million in total, suffered severe operating difficulties for two years during which it was usual for only two of the three to be in operation at any one time (*The Times*, 1969 a). Finally, sustained high levels of operation are necessary to cover the high standing charges and failure of the unit during its short economic life may also be financially disastrous.

Marketing problems are a more obvious but probably less important aspect of giant, concentrated operations. Shipment in bulk, by barge, coastal vessel or pipeline, reduces the significance of distribution costs and for some projects large export outlets make even a remote coastal location viable. In extreme cases, availability of deep water and Development Area status are the only factors focussing major new developments or expansion in Britain rather than inside the Common Market. Fifty to 75 per cent of the output of the proposed Invergordon plant was designed for export and British Petroleum seems likely to devote a larger proportion of the Grangemouth output to foreign sales while Baglan Bay takes over more of the home demand (ECN, 1968 a; 1968 b). Milford Haven is an interesting case. By 1973, the Haven is likely to have four refineries (including a new Amoco one) with about a fifth of the United Kingdom's refinery capacity, but only Gulf will produce petrochemicals.

TABLE 7.5. *United Kingdom: ethylene production capacity 1958, 1964 and 1968 (thousand tons)*

	1958	1964	1968
ICI Wilton	60	140	300 (450, 1969)
Shell Stanlow	25	?	40
Carrington	55	100	250
BP Grangemouth	160	130	400
Baglan Bay	—	55	60 (to expand to 340)
Esso Fawley	40	120	120
British Celanese, Spondon	15	28	30
British Oxygen, Maydown (Northern Ireland)	—	—	15
Total	355	573	1,215

Source: ECN 1965*f*, 21; *Economist*, 1968*e*.

Esso, though it has the oldest and biggest refinery there, soon to be doubled to 15 million tons, keeps its petrochemical operations concentrated at Fawley. The deep water is ideal for supertankers, oil products are distributed to a considerable extent by small coastal tankers, but shipment of chemicals to a home market by drum, bagged or in other small units would involve too much dependence on the indifferent road and rail services. Significantly, Milford Haven is Gulf's first wholly owned petrochemical operation in Europe and its output will have an important export market (ECN, 1965*b*).

The combined effect of increased marketing costs, constructional delays, difficulties at the running-in stage and the vulnerability of the whole operation, when failure of key pieces of equipment may immobilise an investment worth perhaps £20–£30 million, have led some to suggest that scale will not continue its headlong increase of recent years but will tend to stabilise on the bigger units now installed (Caudle, 1969; Arundale, 1969). The ten-year progress of one product, ethylene, is indicated in Table 7.5. It is true that some observers proclaimed an imminent halt to the increase in plant size in the past and were proved wrong, but at least there seems a possibility that any further increase in scale will be less rapid than in the nineteen-sixties. If this is the case and if transport costs continue to rise, some further geographical spread of heavy chemical production may be expected. However, in an industry made up of only a few major firms, much depends on corporate strategies, the different companies reacting in various ways to a given situation, including government policy. The economies and diseconomies of scale, the problems of distribution and the merits of fully integrated operations are well revealed in a major but unspectacular branch of the chemical industry, namely, the manufacture of fertilisers.

A case study – fertiliser production

Fertilisers are of four main types, nitrogenous, phosphatic, potash and compound, that is containing more than one of the others. New preparations have been developed but no major new types have been produced for many years, and the trade differs from much of the chemical industry in that emphasis is on process improvement rather than product development. Assembly of bulk materials and basic chemicals including ammonia and sulphuric, nitric and phosphoric acids is involved and the product is bulky and weighty in relation to value. As a result, distribution costs are high. It is difficult to find accurate figures for freight charges, but some indication of their magnitude may be gained from the estimate that road carriage to Northumberland markets from Merseyside costs about £1 a ton more than bulk rail movements. Under these conditions it is economic to rail fertilisers to distant terminals and then transfer to road vehicles for delivery to the depot in spite of the break of bulk. Notwithstanding these high distribution costs, because processes are capital intensive, each major producer strives for a large share of the national market and cross hauling is common (Monopolies Commission, 1959; Luttrell, 1962; *Chemistry and Industry*, 1969*b*). In the last decade, scale of plant has grown and for most of the time selling prices have fallen and raw material prices increased, so that the market has become more and more competitive. In 1966, 40 per cent of the domestic market was held by Fisons. Although it had already been diversifying into more profitable lines of business for thirty years, two-thirds of the company's capital was still tied up in fertilisers, which in 1965 provided no more than 55 per cent of the turnover and 46 per cent of group profits (ECN, 1966*a*; *Times Review of Industry*, 1966). ICI controlled 25 per cent of the market and Shellstar 15 per cent – at that stage operating from its petrochemical complex at Shell Haven. Other, smaller producers were ground between the big concerns which, in addition to their scale advantages in fertilisers, had other lines to tide them over bad times. One of these smaller companies, the oldest fertiliser firm in Britain, illustrates very well the problems of the small producer.

Sir John Lawes of Rothamsted built works to make super-phosphate of lime at Bow in 1843. In the nineteen-sixties, Lawes Chemical Company, Limited, still turned out straight fertilisers and compounds at its Barking plant, but had only 5 per cent of the national market. Its ammonia supplies were bought from the Becton works of the North Thames Gas Board and from ICI. As a small producer, the firm was badly hit by the cost/price squeeze of the mid-nineteen-sixties. By early 1966, compound fertilisers were selling at £4 a ton below the 1958 level but in the previous eighteen months alone sulphur prices had risen 45 per cent (ECN, 1965 c, 1966 a). Sales by Lawes shrank and in the fifteen months to September 1966 the firm lost £0.5 million (ECN, 1966 b). Attempts were then made to reduce costs. In 1965, a long-term contract was signed with the American firm, W. R. Grace, Inc., for delivery of liquified ammonia from Trinidad to be stored for Lawes on the estuary at Thameshaven. In Trinidad large supplies of natural gas were available to the ammonia producer at an estimated 0.21 pence* per therm as compared with subsequent bulk rate sales of North Sea gas to the chemical firms at 1.7 to 2.1 pence. Lawes followed this agreement by installing a new and highly efficient plant at Barking to double fertiliser capacity to 120,000 tons a year while cutting the labour force by 40 per cent. Delays in commissioning plagued this investment and by the time it was at work in January 1969 prices had fallen to a point at which returns were below its break-even point. Lawes' predicament was that it was too big for merely local marketing – and even in the local market was challenged by a Fison depot next door in Barking – but too small to justify really large, low-cost plant or to get all its materials at favourable rates. In July 1969, the firm went into voluntary liquidation (ECN, 1965 c, 1966 b, 1969 c, 1969 d).

Others too have had to balance the advantages of a wide spread of units against the diseconomies of their smallness, or to opt for a largely local market by choosing to locate in areas where fertiliser consumption is especially intense, or, finally, to develop a highly efficient distribution system for a mass market (Figures 7.1 and 7.2). The new Albright and Wilson plant at Barton on Humber is an example of a project designed for a regional market. Albright and Wilson control only 6 per cent of the United Kingdom compound fertiliser market but the decision was taken in 1964/5 to build a new plant to concentrate the operations formerly spread among five manufacturing units, and to serve mainly the Yorkshire and Lincolnshire farming areas. Even so, the firm has publicly expressed its own doubts about the profitability of its new operation (*The Times*, 1969 b). The 0.33-million-ton liquid fertiliser plant which Esso built at Warboys, Huntingdonshire, in 1967 is a variant on the same theme. The location was chosen because of the high consumption in East Anglia,

---

* In this chapter all values are expressed in decimal currency.

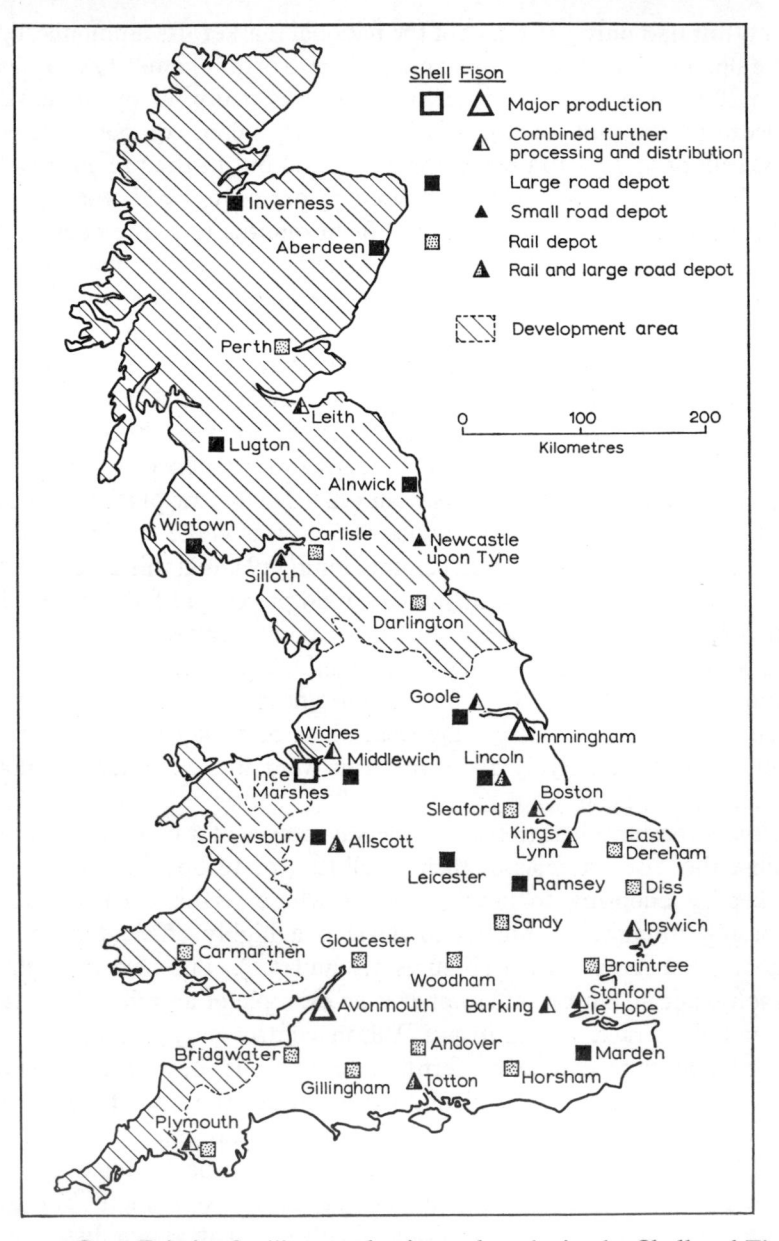

Figure 7.1. Great Britain: fertiliser production and marketing by Shell and Fisons, 1970.
Based on company information.

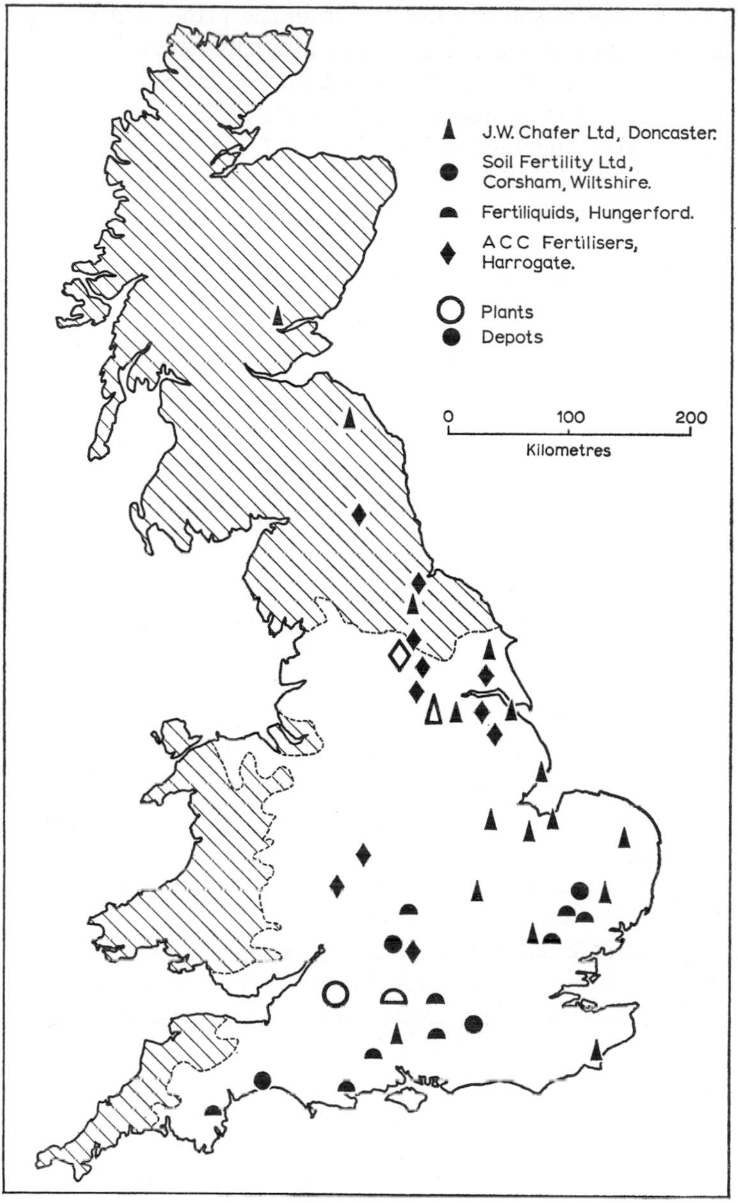

Figure 7.2. Great Britain: fertiliser production and distribution by four smaller firms, 1967/8.
Source: *Farmer's Index, 1967/8.*

though the variety of soil conditions within the plants' marketing radius of 32–48 kilometres (20–30 miles) was also initially stressed as a locational advantage. Late in 1969, when it decided to close the Warboys operation, it was claimed that it had been largely a test production and marketing exercise. Its size suggests it was intended to be rather more than that. Another factor in this failure was the conservatism of the British farmer, which made him reluctant to invest in spray plant (ECN, 1969e).

The bigger producers have arranged to cover the national market more completely. ICI has a broad grip on the major markets of the English lowlands from its units at Billingham, Immingham and Avonmouth. Fisons operated a larger number of producing units when distribution was less easy and when sulphuric acid and superphosphates were the basis of the business. Just before World War II, they had almost twenty plants, not systematically located but in a rather haphazard pattern which reflected their mixed origin, partly built by Fisons, partly acquired by purchase of other smaller concerns. With technical change, Fisons have focussed more and more of the primary stages of production at Avonmouth, where a small site was acquired as early as 1936, and at the much bigger Immingham operation, where production began in 1950. Other units have combined further processing and the function of distribution depots, all of them coastal, as at Leith, Goole, Boston, Kings Lynn, Ipswich, Barking, Stanford le Hope, Plymouth and Widnes. There are further big depots at Lincoln, Totton (Hampshire) and Allscott (Shropshire) and smaller ones at Silloth (Cumberland) and Newcastle upon Tyne (Fisons Limited, 1970). When Shell decided to enter the fertiliser business in a bigger way in the mid-nineteen-sixties, it decided on another strategy. Since 1960, Shell had operated an ammonia, nitric acid and nitro-Shell fertiliser plant at Shell Haven, but now along with the American firm Armour decided to build a much bigger unit. A location on Teesside near to the new Teesport refinery and to the potash field which was soon to be opened was put aside in favour of Ince Marshes near Ellesmere Port. Although this was the first major fertiliser plant in the north west, it was clear that its 0.75 million-ton capacity would need a much wider market. By 1970, when the Ince Marshes operation was, very belatedly, brought into production, Shell had negotiated a major long-term contract with British Rail which fitted into a grand marketing strategy. Company trains will move 400,000 tons of product from Ince Marshes every year in loads of 10 waggons, each containing 48 tons of bagged fertiliser. They will deliver to 24 depots, each having storage capacity for 5,000 to 10,000 tons and located some 80 kilometres apart in the major agricultural areas (ECN, 1970b).

The economies and diseconomies of integration are also illustrated by the fertiliser business. ICI and Shell, with oil feed-stock supplies, natural gas and links with mineral working, have fairly fully self-contained operations. One reason given for the failure of Esso's Warboys operation was that its nearest controlled source of ammonia was at Rotterdam (ECN, 1969e). Fisons is a more complex case. The firm has nitric acid capacity, the largest phosphoric acid plant in Britain and some ammonia capacity. Nevertheless, it has to buy its minerals and also much of its ammonia. This policy exposes it to bigger price swings than its more integrated competitors. In the two years to spring 1966, the world price for bright sulphur ran from under £10 to well over £19 a ton. Later, however, the pendulum swung the other way. In 1966, the company pulled out of potash exploration in north Yorkshire and in 1969, as Shell and ICI poured money into development there, Fisons benefitted from a fall in world potash prices to an all-time low. Early in 1969, the company left the National Sulphuric Acid Association's Sulphur Pool in order to gain full advantage from a fall in prices which it believed might save £0.25 million in a full year. Subsequently, sulphur prices tumbled still further, the export price of Frasch sulphur f.o.b. from US Gulf ports falling in 1969 alone from £16.8 to £8.4 per metric ton (ECN, 1966c, 1967c, 1969f; *Times Review of Industry*, 1966; *Mining Journal*, 1970). In the case of ammonia, ICI had to bear the cost of all the commissioning troubles at Billingham while Fisons had made its purchases at contract prices. However, by 1970 Fisons was known to be worried by its long-term contract for ICI ammonia and was already paying well above current rates of £17 a ton.

The fertiliser business illustrates in extreme form the dilemma of reconciling manufacturing scale economies with high distribution costs. The six major plants operated by the big three producers represent a very large capital outlay but have a relatively small labour force, and only one, ICI Billingham, is located in a Development Area. In contrast with the fluids and gases of other branches of the chemical industry, their product is not easily transported. Public policy must be concerned with the relation between national and company interest in the matter of plant location and product distribution.

Competition in fertilisers is keen. To obtain the business which alone justifies large scale operations, the major producers compete throughout the country. It is common for one firm to cross haul into the 'natural' market area of a competitor. If this carriage is by rail, although British

Railways secures a welcome revenue, there may be a misallocation of scarce national resources. If it is by road, social costs are added to company costs, although they are less easily measured.

The chemical companies are now beginning to coordinate their major investment programmes. Perhaps some marketing rationalisation might also be possible, with each plant supplying the nearest markets, or, where two plants are in one location, as at both Avonmouth and Immingham, dividing the market area. True, farmers already complain about oligopoly in fertilisers, but efficient quasi-official oversight might ensure that social and company costs were both reduced on the distribution account while preserving scale economies and the rights of the farmer. Even with rationalisation in marketing, it would be possible to avoid the elimination of competition, for market areas would widen or narrow with changes in the relative costs of production at each of the major plants.

The chemical complex

The existence of services for one chemical process or plant helps to attract other developments to the same site and the availability of the products of one firm may well attract major further processing. This helps to create big agglomerations of capacity, but new transport techniques make it possible for these complexes to be very widely spread geographically, a trend which creates problems for the regional planner looking for major agglomeration effects from a heavy chemical development.

The scale economies considered earlier seem in some cases to have caused a polarisation of developments with bulk exchanges taking place between specialist areas. The speculation in 1965 whether the new ICI ethylene capacity would be built at Runcorn rather than Wilton was resolved in favour of the latter. However, in the same year a 225 kilometre (153 mile) Trans-Pennine ethylene pipeline connecting the Tees with the Mersey was announced at a cost of as little as £2 million, and as a result ethylene capacity and expansion have been focussed at Wilton with low cost delivery from there of the ethylene requirements of ICI's north western plants. In turn, the latter has become the focus for major expansion of that company's chlorine capacity, with two 200,000-ton units announced within one year (ECN, 1965a, 1965d; *Chemistry and Industry*, 1965, 1966). The wider significance may be seen in the start of what may be a major British ethylene grid. The first element in this has been announced – a 6 kilometre (4 mile) link from the ICI Trans-Pennine pipeline to Shell's Stanlow–Carrington line. The link will have an annual capacity of 120,000 tons and will help to level out both company's surpluses or deficits. A wider grid including Baglan Bay, Grangemouth and Salt End (Hull) has been suggested (ECN, 1968 *a*). There are other examples; for instance, the new North Teesside aromatics plant which will supply 80,000 tons a year for production of cumene (a hydrocarbon) at Grangemouth, most of which will in turn be shipped back to Billingham (ECN, 1965 *e*). New bulk transport methods and especially pipeline transport, taken in conjunction with the scale economies desirable today, seem capable of making the whole of industrial Britain into one interlinked chemical complex. If this is so, then the concept of the MIDAS (Maritime Industrial Development Area) is too narrow (McKitterick, 1970). Contemporary techniques in the bulk movement of fluids and gases place all the industrial areas of Britain within the supply radius of any major deepwater chemical complex. As scale economies grow beyond the capacity of national markets to support them, so even wider scales of international linkage are emerging.

Tankers already carry up to 100,000 tons of Teesside ethylene annually to Rozenburg, and Trinidad sells 500,000 tons of ammonia to the United States and European firms. A spectacular example which illustrates the pitfalls as well as the gains from an intra-company variant of this pattern is provided by the largely inorganic chemicals firm of Albright and Wilson. With wide interests in fertilisers and other heavy chemicals, Albright and Wilson's have also in their Oldbury Division a large investment in phosphorus and have operated two major United Kingdom phosphorus plants at Oldbury and Portishead. In the mid-nineteen-sixties, cost analysis suggested that great savings might be made if the primary processing were undertaken in Newfoundland with further processing only retained in Britain. A new subsidiary, ERCO (Electric Reduction Company), was set up to build a plant at Long Harbour on the southern coast of Newfoundland. It was estimated that the costs for the delivery of phosphorus in the United Kingdom might be as much as 18 per cent below the home production costs (Table 7.6). Power would be very cheap, Florida phosphate rock could be delivered there at much below the English price of Moroccan phosphates and, as labour and capital charges represent 30 per cent of production costs, concentration on one big and efficient unit would give additional big savings. The two British plants would close when ERCO was at full production and shipping half its output across the Atlantic. By spring 1969, an investment of £1.2 million in old British phosphorus plant had been written off. However, it then became known that the Newfoundland plant would cost 10 per cent more than had been estimated and take three months longer to complete, with all the adverse implications for

TABLE 7.6. *United Kingdom and Newfoundland: estimated costs of production for phosphorus, circa 1968 (£ per ton)*

	United Kingdom (45,000 ton capacity plant)	New-foundland (90,000 ton capacity plant)
Raw materials	47	38
Electricity	40	20
Labour, maintenance and overheads	7	6
Depreciation and interest at 15 %	23	19
Total production cost	117	83
Cost of transport to UK	—	8 (later 4)
Terminal charges	—	4
Total costs	117	95

Source: ECN, 1968 d.

the return on the investment considered above. Soon after this, the ERCO plant was closed as a result of complaints to the provincial government that its effluent was killing the fish in Placentia Bay. Standing idle, the plant was a £50,000-a-day liability, £0.5 million more had to be invested on an effluent treatment plant and meanwhile £1.5 million was spent to restart the Portishead furnaces, in keeping those at Oldbury at work and in buying phosphorus. Conditions later improved, but the pitfalls in reaping the advantages of international scales of operation had been effectively shown (ECN, 1968 d and *passim*).

Despite such problems there remains the probability that much more of future development in heavy chemicals will be internationally based, and in some cases on the Albright and Wilson model. Such a likelihood poses still more problems for the regional economic planner.

The impact of natural gas on the chemical trade

The concentration of the chemical industry on continuous production processes using gases and fluids seems ideally fitted for natural gas, and fortunately two of the major agglomerations of heavy chemical plant, Teesside and Humberside, are near to the North Sea fields. As late as 1966, it was possible to make ammonia more cheaply from naphtha than from natural gas at the prices then being negotiated – 2.1 pence per therm

TABLE 7.7. *United Kingdom: estimated costs of production for ammonia in a plant of 330,000 tons capacity, circa 1969 (£ per ton of ammonia)*

	Naphtha (£10 a ton)	Natural gas (1.7 pence per therm)	Natural gas (1.9 pence per therm)
Raw materials and feedstock	9.1	6.3	7.0
Utilities, labour and maintenance	1.5	1.5	1.5
Capital charges at 16.5 %	2.7	2.4	2.4
Total costs	13.3	10.2	10.9

Source: based on ECN, 1968e, 28; 1969g, 4.

in the Gas Council's initial agreement with British Petroleum. Thereafter, partly because of the closure of the Suez Canal and partly because of devaluation, naphtha prices rose and at the same time it became clear that it was reasonable to look to lower natural gas prices.

By 1968, ammonia from plants in the Americas could be delivered in the United Kingdom for £12.2 per ton and from the Netherlands at an estimated £10.6 per ton. Big United Kingdom ammonia producers had to find a feedstock to enable them to meet these prices. Naphtha could no longer do this but closely negotiated contracts for North Sea gas offered new hope, and in 1969 agreements for the supply of 1.7 million cubic metres daily to Shell and ICI were announced. Shell will pay 1.9 pence per therm; ICI a variety of prices ranging down to 1.7 pence. ICI obtained a lower price partly because Billingham will take gas for a 900,000-ton ammonia unit, as opposed to only 350,000 tons at Ince Marshes, and partly because its Teesside works are only a short distance from the Gas Council pipe. Not only are £2 to £3 per ton saved on ammonia costs but the scarce naphtha is spared for the carbon-based petrochemical operations which natural gas cannot support (Table 7.7). The availability of bulk supplies of gas is likely to reinforce a tendency for chemical works to cluster into complexes.

The distribution of chemical manufacture in Britain and regional development policy

By no means as much of the chemical industry is located in Development Areas as is the case with some other basic trades, notably steel, and above all shipbuilding. After the great increase in government assistance in the nineteen-sixties, a major impact on the location of growth for such a capital intensive industry might reasonably have been expected. The evidence is however inconclusive. As a general rule, it seems that the primary force in shaping decisions on the location of expansion is pre-existing major plant, whether inside a Development Area or not. However, when decision wavers between expansion at one of a number of existing plants, Development Area status may often be vital. This is especially so in recent years, as the size of capital outlays has increased and as Development Area assistance of all types multiplied six-fold in the four years to 1968/9.

Twenty years ago, the pull of the problem areas was less than now. Fisons were looking for a site for a big new east coast concentrated-fertiliser plant in 1945/6. The company favoured the Humber, which offered both physical and marketing advantages. Board of Trade pressure induced the firm to consider the possibilities of Teesside development. At that time, however, the available Tees estuary sites were judged to lack both sufficient room for disposal of solid waste and adequate dock facilities (Luttrell, 1962, 647). Such a conclusion seems strange in view of the present characteristics of that estuary. The Managing Director of Esso Chemicals has revealed that his firm has contemplated big developments on new sites but at no time has it seemed economic to break away from Fawley in spite of its lack of any special government assistance (*Chemistry and Industry*, 1969c). Early in 1970, Shell announced its £225 million expansion for the north west. Stanlow, within the Merseyside Development Area, is allocated £75 million for refinery expansion and £30 million for chemicals but the rest will be spent at Carrington, 37 kilometres (23m) to the north east and outside the Development Area (*The Times*, 1970a). With a building grant of 25 per cent for Development Areas, and investment grants for machinery and equipment of 40 per cent as compared with 20 per cent elsewhere, the choice of Carrington must have involved increased costs of the order of £25 million. The size of the figure is an index of the attractiveness of expansion at existing plants with their established services, labour force and products, or their need for rounding out. Expected process cost savings more than compensate for the extra capital cost, and Shell indeed have emphasised that the locations for the expansion were chosen purely for commercial reasons and without regard to the grant situation.

The development of ICI at Severnside however points to some deficiencies of a good location and site away from a major Development Area chemical complex. In the mid-nineteen-fifties, it was decided to build a new multi-divisional complex on the Wilton model. Meanwhile, the Agricultural Division, which figured largely in the development thinking, had found that the centre of gravity of British fertiliser demand was in the neighbourhood of Oxford, far from the main east coast ICI concentrations. In 1957, the firm was granted planning permission for development on 400 hectares (1,100 acres) of farmland some 6 kilometres north of Avonmouth. At that time, it was widely expected that £100 million would be spent there by the mid-nineteen-seventies, ICI observing 'The Company hopes to develop this site on similar lines to Wilton Works' (Imperial Chemical Industries, 1957; Walker, 1965; *Petroleum Press Service*, 1957). Two years later, when Wilton's third naphtha cracker was opened, bringing ethylene capacity there to 110,000 tons, it was said that if a fourth unit were needed it would not be at Wilton but at Avonmouth. However, big new ICI developments have been judged against the attractive costs of Wilton expansion and, as capital costs and government assistance have both increased, Avonmouth has been notably absent from the list of major investments (Luttrell, 1962, 631, 634). Wilton ethylene capacity is now 450,000 tons, but there is still no ethylene capacity at Avonmouth, even though its purchases of piped ethylene from Fawley are said to be very costly (*Manchester Guardian*, 1959; ECN, 1968 c). On the ammonia account, Avonmouth capacity is 280,000 tons whereas that of Immingham is 190,000 tons and that of Billingham 900,000 tons (ECN, 1969 g). If the government assistance available for Development Area expansion is abandoned or reduced, it seems likely that ICI will locate more of the next round of ammonia expansion at the first two plants. Presumably also if competition in fertilisers became keener it would give serious attention to the proposals for a new fertiliser complex to serve the prime market of East Anglia and the Home Counties, perhaps involving a Thamesside plant.

The capital intensiveness of the heavy chemical trade means that labour forces do not increase in proportion to output and indeed the need to rationalise in order to meet increasingly keen competition may often mean that labour forces shrink. This is especially the case in the older industrial areas which have a more than proportionate share of the primary end of the chemical trades – so that, for instance, the value per ton of chemicals exported from the South East region in 1964 was 2.5 times that from the North West (ECN, 1966 d; see Tables 7.8 and 7.9). Big Development Area grants have assisted the progress to giant plant but with little new employment, and the Regional Employment Premium has been insufficent to compensate. Other much less impressive developments in other industries may yield more jobs. For a wide range of light to medium engineering it is reckoned that the average total cost of plant and equipment per worker

TABLE 7.8. *United Kingdom: employment in chemicals and other industries, by standard regions, 1963 (percentage of United Kingdom totals)*

Standard regions	Chemicals (order IV)	Metal manufactures (order V)	Ships and marine engineering (order VII)	Vehicles (order VIII)	All manufacturing
Northern	12.8	9.0	19.8	1.6	5.0
North West	24.2	6.4	14.4	14.8	15.9
Wales	5.1	15.1	—	—	3.5
Scotland	7.0	8.2	21.6	4.1	8.0
N. Ireland	0.5	0.1	7.3	3.1	2.1
Yorkshire and Humberside	9.7	19.6	3.5	4.8	10.3
South East	28.7	8.5	22.0	30.6	28.0
West Midlands	4.3	24.8	1.0	25.4	14.0
Others[a]	7.7	8.3	10.4	15.6	13.2
Total	100.0	100.0	100.0	100.0	100.0

[a] East Anglia, East Midlands and South West.
Source: *Census of Production*, 1963, *General Tables* (Board of Trade, 1969).

is £2,300 to £2,500, with a similar figure for construction costs, or an overall investment per employee of about £5,000 (NEDC, 1970*b*). Shell's £225 million investment in the north-west will attract government financial assistance of approximately £35 million but the increase in the labour force there will be no more than about 1,200, whose earnings will inject only £2 million annually into the regional economy (*The Times*, 1970*a*).

Teesside points up the problems especially well. Monsanto is investing £10 million in a completely new Teesside plant but its work force will be only about 100, that is an investment per worker of £100,000 including possibly £35,000 of government assistance. Yet the average cost to the government of each new job created in the North East in 1969/70 was only £640 (NEDC, 1970*a*). ICI spent £24 million at Wilton in 1968/9 but with little or no increase in its labourforce. At the end of 1969, a £3 million sulphur burning plant was announced for Billingham. It will replace four older sulphuric acid units. The labourforce will be cut by 250, including 100 from the anhydrite mines where output will fall by a third, though the company has expressed a hope that they will be redeployed (*Chemistry and Industry*, 1969*d*). *Teesplan* (the Teesside Survey and Plan, 1969) has

TABLE 7.9. *United Kingdom: net output of chemicals and other industries by standard regions, 1963 (percentage of United Kingdom totals)*

	Chemicals (order IV)	Metal manufactures (order V)	Ships and marine engineering (order VII)	Vehicles (order VIII)	All manufacturing
Northern	14.3	8.2	19.5	0.8	5.2
North West	25.8	6.0	16.1	13.2	15.3
Wales	4.2	17.9	—	—	4.0
Scotland	7.0	7.9	19.6	4.0	7.8
N. Ireland	0.7	—	6.5	2.6	1.6
Yorkshire and Humberside	8.0	18.6	3.3	4.0	9.4
South East	30.0	8.6	23.3	36.0	30.7
West Midlands	3.4	24.2	1.5	25.7	13.4
Others[a]	6.6	8.6	10.2	13.7	12.6
Total	100.0	100.0	100.0	100.0	100.0

[a] East Anglia, East Midlands and South West.
Source: *Census of Production*, 1963, *General Tables* (Board of Trade, 1969).

forecast that between 1966 and 1991 employment in the area's chemical industry will fall from 33.1 to 31.6 thousand.

All this suggests scant return for large government financial help. The situation is the inevitable result of the divergence in the geographical scale at which the processes are organised and the area from which a complex can draw its workers. It is suggested that modern bulk transport makes the nation one heavy industrial complex, but each industrial area within that complex has its own distinctive labour catchment area. In an area preoccupied with the primary end of the business, such as Teesside, employment suffers the full effect of technical progress as firms rationalise to meet the competition of evermore efficient foreign production. Bray (1970) has suggested that existing Development Area programmes do not get to the heart of the regional problem. Pointing out that investment help to the Northern region's chemical industry has enabled it to cut rather than increase its labour force, he suggests that the relative neglect of lighter trades, service employment and higher education have left 'an unbalanced and unstable local economy'. This opinion may be said to pay inadequate attention to the competitive pressures in heavy chemicals, and there are two important qualifications to be made.

First, a static or even a shrinking chemical labourforce does not rule out growing overall subregional income from this industry. In 1969, ICI paid £40 million in wages and salaries on Teesside and one quarter of all rates of the new Teesside County Borough (*Teesside Journal of Commerce*, 1969a). These outgoings will support the real growth sector of the previously unbalanced Teesside economy, services, whose labourforce is expected to grow from 97,000 to 164,000 over the same twenty-five-year period analysed by the Teesside Survey and Plan (1969). Heavy government investment in economic activities which seem to do nothing directly to solve the high unemployment problem is in fact justifying a new surge in the largely privately financed tertiary sector.

Second, the whole development might snowball. Stage one of this process may involve an agglomeration of firms brought to an area by common calls on the environment or the labourforce or by the need for each others' products. For many years Billingham and Wilton operated in the company of only a few much smaller operations such as those of British Titan Products. Then came the Phillips–Imperial refinery and Shell at Teesport. More recently, three American firms have built or announced important works for Seal Sands – Monsanto, Lennig an acrylate monomer and methyl methacrylate plant, and W. R. Grace a sodium nitrilotriacetate operation. Nearby at Greatham, British Titan Products is building a second Teesside works. Lennig chose the Tees because of the abundance of space and proximity to its existing Jarrow works, Grace because there are firms there which could provide its raw materials. In spite of the wide spread of heavy industry linkages there have been suggestions of a further, more advanced stage of complex building on Teesside. It is said that eventually polymer based products will probably replace conventional materials in the car, furniture and building industries and hopes have been expressed that therefore some firms in these industries may wish to build on Teesside (*Chemistry and Industry*, 1969e). Whatever the implications for the last two trades, it is clear that the agglomerative links of the motor trade are with its major concentrations of component firms and these connections will be less easily broken or stretched than in 1960/1, when buoyancy and expansion in motors was a more prominent theme than now. Though Teesside has locally available steel and plastics raw materials, its present emphasis even within these trades ill equips it as a base for the motor trade.

One may conclude that even though the hopes for a diversified Teesside industrial complex have often been too optimistic, the state of its heavy chemical trade points to the need for continuing major investment assistance. Without this, expansion will probably drift away to growth areas nearer to the centre of gravity of the national market. More damaging still to the economy, there seems a likelihood that some of this growth, and perhaps a great deal of it, would be siphoned off to locations within the European Common Market.

Physical planning and chemicals

Chemical manufacture presents serious problems to a society increasingly conscious of the quality of its environment, and as the economic gains from agglomerations of capacity increase so also do the physical planning difficulties. Unobtrusive underground mines or pumping operations supply some of its mineral needs but in other cases these involve open pits. The great limestone quarries which supply plants in the Mersey district circle Buxton and although some have weathered into congruity with a natural landscape of fine limestone scars others are still noisy, drenching the hillside with white dust and causing occasional dreadful sludges. In Cleveland, the opening of the potash deposits is bringing not only highly valued new employment but also controversial mine headgear and processing plant into a National Park. Interworks transfers bring slow tanker vehicles onto the main roads laden with cargoes whose toxicity is hidden under a host of awe-inspiring chemical names. Processing plant requires large areas of flat ground, land which is therefore frequently of good agricultural quality. From the impressive collection of fractioning columns, pressure vessels, domes, tanks, cooling-towers, chimneys and the convoluted pipework which links them, steam, multi-coloured gases and a range of odours escape at point after point. The movement of waste into the atmosphere is matched by that which finds its way by various sluices into streams or estuaries. Tall chimneys along the Manchester Ship Canal and the Mersey pour out their waste over the low ground. Haze or thickly charged mists drift high over industrial Teesside from Billingham and, nearer at hand, Haverton Hill and Port Clarence huddle on its lee side as filthy, dust-strewn communities. Teesside is a broad estuarine lowland with a sharp escarpment to the south and more gently rising ground on its north side, so that when there is a breeze from the east or a temperature inversion in wintry conditions the build-up of pollution can be especially dangerous.

One hundred years ago, apologists for the alkali trade attempted to show that neighbourhoods over which their white clouds of hydrochloric acid gas rolled were healthier as a result. Attempts are no longer made to match these early exercises in the manipulation of vital statistics. Chemical companies and communities are well aware of the problems and increasingly willing to tackle them (Shell, 1969). Over the last ten years, ICI have spent £28 million on effluent control throughout Britain and expect to step this up to £60 million over the next ten years (*Teesside Journal of Commerce*, 1969b). *Teesplan* makes provision for heavy industry to be still more carefully zoned at the seaward end of the estuary away from the bigger centres of population. Heavy industry will always create nuisance, and therefore the burden of planning should be to reduce this to a practicable minimum and to find sites which provide the best combination of operating conditions with the minimum incompatability with neighbouring land uses. It must be admitted that in an industry where locational inertia is as prominent as in heavy chemicals this must frequently appear as a counsel of unattainable perfection.

Some of the notorious heavy chemical districts of the past have disappeared. The remnants of the old Tyneside alkali industry are still an unsightly if intriguing mess except where in 1968 they were landscaped at the start of a project for a riverside recreation area. The site of the St Rollox works now sprouts impressive multi-storey flats rehousing people from the teeming tenements of east Glasgow. The Midland Link motorway carves its way through the sulphurous wastes of Oldbury. Yet today's great Merseyside complex can be traced back not only a century to Brunner and Mond's introduction of the Solvay Process to Winnington but to Muspratt's arrival with the Leblanc process in Liverpool fifty years before. Brunner and Mond already owned some of the land in the Billingham area by the late eighteen-nineties. The smaller limestone quarries of Derbyshire and the northern Pennines supplied the needs of Lancashire and north eastern chemical works in the Victorian age. Though not in the extreme form of steel or ships, even the British chemical industry, for all its growth and chameleon-like changes of materials, processes and products, is strongly conditioned by its past both in its geographical pattern and the planning problems which it causes.

Conclusion

Scale economies in heavy chemicals make centralised operations desirable; bulk transport makes them practicable. The biggest existing primary plants are very highly capitalised and can increase their output with a stable or even a decreasing workforce. Finishing operations, frequently much more labour intensive, can be on a smaller scale and more widespread. Therefore, the existence of heavy chemical plants does not imply an automatic local economic growth spiral. Despite their visual impact and large investment outlay, heavy chemicals are not ideal industries from the point of view of the regional planner. Indeed, as they cause extensive pollution and nuisance, they may discourage other, independent industries. On the other hand, they form a major element in employment in British Development Areas. Increasingly keen foreign competition makes it essential that they be provided with every incentive to modernise and that port facilities, plant sites and other aspects of infrastructure be made available in locations attractive to the industry. Relatively few parts of Britain offer the necessary range of conditions and planners at national, regional and local level must do all in their power to make the desirable facilities available, allocating priorities, reconciling conflicting land uses and so on. Such a task is obviously not easy, but unless it is attempted home growth in British heavy chemicals may slow down if more expansion is transferred to old or new plants across the narrow seas in the heartland of the Common Market. Perhaps the cardinal point to note is the inherent limitation on deliberate government policy in respect of an industry that is, more than most, subject to a double location constraint: the small number of sites within the United Kingdom that are suitable; the intense international competition that characterises the industry. It may well be, therefore, that blanket regional policies covering all industry are particularly inappropriate in this case and that there ought to be greater selectivity in the formulation and application of regional policies.

References

Arundale, D. G. (1969). 'Still bigger plants bring risks', *The Times*, 11 September.

Bell, D. M. (1967). 'The British chemical industry and continental Europe', *Chemistry and Industry*, 6 May, 741.

Board of Trade (1969). *Report on the Census of Production 1963*, CXXXI, *General Tables*.

Booth, N. (1969). 'The ethylene boom', *Chemistry and Industry*, 25 August, 1544–5.

Bray, J. (1970). *Decision in Government*, Gollancz.

Caudle, P. G. (1969). 'Petrochemicals and polymers in the next decade', *European Chemical News*, 28 November, 10.

Chambers, P. (1966). In ICI, 1966, 14.

Chambers, P. (1968). 'The investment problem of the chemical industry', *Chemistry and Industry*, 3 August, 1012–17.

Chemicals EDC (1967). *Manpower in the Chemical Industry: A Comparison of British and American Practice*, HMSO.

Chemicals EDC (1970). *Economic Assessment to 1972*, HMSO.

Chemistry and Industry (1965). 30 July, 18; (1966). 3 June, 23; (1967). 6 May, 745; (1969a). 26 July, 995; (1969b). 22 November, 1674; (1969c). 30 August, 1191; (1969d). 20 December, 1815; (1969e). 7 June, 735.

Economic Research Group (1967) (of a group of European Banks). *The Chemical Industry in some European Countries.*

Economist (1966a). 1 January; (1966b). 19 November, 828; (1967). 7 October, 63; (1968a). 2 March; (1968b). 27 July, 64; (1968c). 11 May, 79; (1968d). 17 February, 72; (1968e). 28 December; (1969). 24 May, 73.

European Chemical News (ECN) (1965a). 9 August, 20; (1965b). 28 May; (1965c). 10 December, 9; (1965d). 1 October, 22; (1965e). 10 September, 4; (1965f). 26 February; (1966a). 11 February, 8; (1966b). 30 December, 32; (1966c). 3 June, 8; (1966d). 22 July, 10; (1967a). 13 October, 6; (1967b). 29 September, 5; (1967c). 27 October, 30; (1968a). 14 July, 28; (1968b). 1 November, 8; (1968c). 26 July, 22; (1968d). 5 July, 30; (1968e). 30 August; (1969a). 28 November; (1969b). 4 April, 14; (1969c). 20 June, 52; (1969d). 25 July, 69; (1969e). 21 January, 6; (1969f). 12 December, 26; (1969g). 17 January, 4; (1970a). 16 January, 18; (1970b). 3 April, 12.

Fenning, R. M. F. (1968). *Financial Times,* 29 July, 24.

Fisons Limited (1970). Personal communication.

Hardie, D. W. F. and J. D. Pratt (1966). *A History of the Modern British Chemical Industry,* Pergamon.

Imperial Chemical Industries (1957). *Annual Report,* ICI.

Imperial Chemical Industries (1966). *Annual Report,* ICI.

Imperial Chemical Industries (1968). *Annual Report,* ICI.

Iliff, N. (1969). 'New horizons for chemistry and industry in the 1960s', *Chemistry and Industry,* 9 August, 1073–6.

Luttrell, W. F. (1962). *Factory Location and Industrial Movement,* National Institute of Economic and Social Research, 2 vols.

McKitterick, T. E. M. (1970). 'Midas revisited'. Paper read to the Maritime Economists Group, 12 August.

Manchester Guardian (1959). 16 September.

Mining Journal (1970). *Annual Review,* 96, 98.

Monopolies Commission (1959). *Report on the Supply of Chemical Fertilisers,* HMSO.

North East Development Council (1970a). *Ninth Annual Report, 1969–1970,* NEDC, 6.

North East Development Council (1970b). Personal communication.

OECD (1970). *The Chemical Industry 1968/1969,* OECD.

Petroleum Information Bureau (1969). *The United Kingdom Refining Industry,* PIB, 2.

Petroleum Press Service (1957). August, 306.

Shell (1969). *Environmental Conservation,* Shell.

Shell (1970). *Annual Report,* Shell.

Teesside Journal of Commerce (1969a). January, 6; (1969b). September, 204.

Teesside Survey and Plan (1969). *Teesplan,* HMSO.

The Times (1969a). 17 July; (1969b). 27 June; (1970a). 16 January; (1970b). 26 February; (1970c). 27 May.

Times Review of Industry (1966). March, 19.

Walker, F. (1965). 'Economic growth on Severnside', *Transactions and Papers.* Institute of British Geographers, xxxvii, 5–6.

8. Freight transport costs, industrial location and regional development

MICHAEL CHISHOLM

The national income accounts show that in 1969 the transport sector, excluding communications, contributed 6.2 per cent of the Gross Domestic Product of the United Kingdom. Back in 1956, the figure was 6.9 per cent and in the intervening years the proportion, though declining slowly, has been very stable. Edwards (1970a), using census data for the production and distribution industries, concluded that 'it is probable that transport accounts for at least 9 per cent of the total cost of producing and distributing [goods]'. On the face of it, spatial variations in the incidence of transport costs ought to be a highly significant factor for industrial location and for regional development generally.

Yet, in his review of the available post-war literature on industrial location in this country, Brown (1969, 778) came to the following conclusion:

The trend of thought in this field has been towards the realisation that transport costs are of relatively minor importance in the majority – and an increasing majority – of industries, that adequate supplies of trainable labour (for some purposes, and for some firms, already-trained labour) are of paramount importance in the post-war situation of relatively full employment, that managerial communications with clients, suppliers, sub-contractors, colleagues and various professional services loom large, and that amenities are important – these last two especially to the people who make the locational decisions.

Brown could have added that it is widely accepted that though transport costs may be a negligible factor in industrial location, other attributes of the transport sector are important; speed, reliability, convenience, packaging problems and the feasibility of integrating transport services into the production process are all given serious consideration by industrialists.

The concept of transport costs is both imprecise and tainted with controversy. For the purpose of this chapter, it is taken broadly to mean expenditure on the movement of goods between points of production and/or consumption. Such expenditure normally includes the purchase of

[213]

transport services or outlays on own-account vehicle operations but excludes costs of packaging, insurance and the costs associated with modifications to production schedules and production facilities. The concept used here is therefore incomplete, largely on account of the lack of relevant data to provide magnitudes for the more comprehensive view of transport costs. Furthermore, the view taken in this chapter is essentially a macro-view and is not concerned with the micro-locational problem of intra-regional spatial allocation. The geographical scale is pitched in terms of the standard, or planning, regions, of which there are ten covering Great Britain; it is the effects of inter-regional location that will be considered. Thus, the central issue may be conceived in terms of an increment of population (of say one or two million persons) for whom a location must be found but who will then generate demands for freight and associated transport services. For the purpose of analysis, the unrealistic assumption is made that all of the additional people would be located in or near major existing urban centres or groups of centres, with an aggregate existing population of at least half a million, or housed in brand new urban developments which achieve that sort of threshold. The question to be examined may then be phrased as: 'To what extent, if any, will inter-regional location alter the demand for freight transport?', taking a long-term view and in a context of government having to make strategic location decisions.

Formal location theory in the tradition of Weber (1929), Lösch (1954) and Isard (1956) leads one to expect that regional differences in location would have a significant impact on industrial costs. The arguments of Myrdal (1957) and Caesar (1964), reinforced by notions of economic potential (Clark, 1966; Clark, Wilson and Bradley, 1969) lead to the expectation that cumulative development in central areas is to be expected, to the detriment of the more peripheral regions. Such cumulative development is seen to operate through the benefit of external economies, of which one is the reduction in transport needs arising from the proximity of firms. The contrary view that diseconomies of further concentration will automatically lead to dispersal is not popular; many observers opine that private considerations favour yet further concentration while the public balance-sheet indicates the desirability of dispersal from the major centres of population. But the necessary empirical evidence is scarce.

To proceed with the analysis, it is necessary to examine a little more closely the conceptual framework that may be employed and it is to this question that the next section is addressed.

The framework for discussion

Public and private costs and benefits

In the post-war period, the greater part of government assistance to the development areas/districts has been in the form of financial inducements to individual firms, either to move them into the aided areas or to expand their activities there. Given this focus of government intervention, it is not surprising that many enquiries have examined questions of industrial location and regional development from the viewpoint of the firms and the questions that weigh with them in making their decisions. As already noted (Brown, 1969), numerous studies of industrial firms have shown that transport costs seem to be a comparatively unimportant factor in their location decisions.

However, while there is a fair degree of unanimity among observers on the evidence as reported, there is considerable confusion concerning the interpretation to be placed thereon. On the one hand, enquiry of firms can only reveal the magnitude of variation in private costs and benefits, whether real or perceived; it does not provide evidence relating to the balance of advantages and disadvantages for the public at large, that is evidence relating to national welfare considerations. If firms are asked to compare their operating costs and profitability in two or more locations, this means one of three kinds of comparison:

1. Between two plants engaged in similar activities but situated in two different locations.
2. Between a location that has been vacated and another that has been occupied.
3. Between a location that actually is used and one or more hypothetical locations that might be occupied.

The major study concerned with the first type of situation (Luttrell, 1962) took a sample of firms that had established branch plants at some distance from the parent plant; the time-period for which data were collected was confined to a relatively few years, of the order of three to four. Most studies have used information under headings 2 and 3 above, especially the third (see Scottish Council, 1962, for example). Almost by definition this kind of comparison is apt to be short-term in character and unable to take account of developments in the distant future.

The distinction between private and welfare considerations and between the short- and long-term view may be conceived as in Figure 8.1. The greater part of the evidence contained in empirical studies to date may fairly

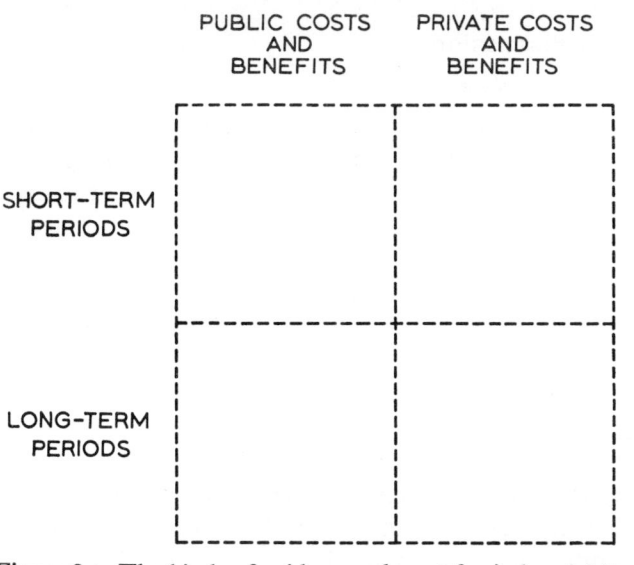

Figure 8.1. The kinds of evidence relevant for industrial location decisions.

be described as falling in the upper right hand sector of the diagram. On the other hand, government policy ought properly to be based on evidence that fits into the lower left hand quadrant. In short, conclusions relevant for one of the four boxes cannot be assumed to be true of any other. *Prima facie*, therefore, one must be wary of drawing policy conclusions from the evidence that transport costs are unimportant for individual industrial firms in their location choices.

The situation is further complicated by the fact that inquiry of firms elicits only part of the picture regarding the incidence of transport costs. It is a widespread practice among British firms to quote national prices delivered c.i.f. to the customer (Chisholm, 1970). One implication of this fact has been clearly recognised by Cameron and Reid (1966): the location of a firm, and especially changes in its location, makes no difference to the delivered price it must pay for manufactured components. It is the suppliers who will have to absorb any increase in transport costs. Therefore, the firm in question is *aware* only of variations in the cost of transport in marketing its own production and may ignore approximately half of the variation in manufacturing transport costs associated with its location choice. Consequently, when firms are asked about the spatial variation in transport costs, they are likely to under-rate its importance by about one half even though giving a correct factual response on the situation as they see it.

To set against this under-reporting of the true position, the relatively short-term basis on which firms can assess the situation implies that they may be inclined to over-estimate the long-term costs of a new location. The reason for this is that, with the passage of time, a firm will find substitute supplies and markets, will reorganise its mode of operation and in other ways adjust to the new location. In all probability, the effect of these changes will be to reduce costs and increase revenue.

Whether this over-estimation of costs will exactly compensate for the under-reporting already noted is a question that could only be answered by empirical inquiry of a kind that has not been undertaken in this country. Suffice it to note that there is a substantial area of uncertainty, which is compounded by a factor not yet mentioned. The majority of inquiries have been conducted among firms actually located in particular places, for example in Scotland and in London. Consequently, if firms report substantially similar structures of operating costs and are not aware of serious transport disabilities, this may be interpreted as evidence of self-selection among the firms in their locational choices, i.e. as reflecting local comparative advantages. If this view is accepted, no conclusions can be drawn as to the effect of changes in location upon the performance of firms. In this context, therefore, Cameron and Reid's enquiry among firms that rejected Scotland as a location is especially interesting:

Of all the factors cited against a Scottish location, geographical inaccessibility was mentioned most frequently. Certainly on the evidence from these seventeen companies, the views expressed by Needleman [and Scott, 1964] and in the Toothill Report [Scottish Council, 1962] seem to be wide of the mark...Many of the seventeen companies concerned were critical of Scotland's geographical position not only on grounds of transport costs but also on the grounds that company *revenue* might be affected (Cameron and Reid, 1966, 23–4).

There is a final problem to which attention must be drawn. Firms are reluctant to disclose their profits and therefore are usually unwilling to make available full details of costs and revenues. The impact of transport is therefore usually assessed in terms of the relationship between transport costs and a company's gross output, net output or total costs of production. On this basis transport costs frequently appear relatively unimportant. However, variations in any one item of costs, including transport, may have a dramatic impact on the 'residual' item, profits. This can be seen from Table 8.1, which shows that when certain outlays specified in the census of production industries are deducted from net output, the 'residual' amount available for all other expenditures (rates, taxes, etc.) and profits is only £2,526 million, or 23 per cent of the total net output. Total expendi-

TABLE 8.1. *United Kingdom: payments made from net output
by manufacturing industries, 1963 (£ million)*

Wages and salaries	5,712	
Capital expenditure	1,025	
Payments for certain services[a]	914	
Transport expenditures	627	
Insurance and pension contributions	388	
		8,666
Remainder of net output		
(profits, taxes, etc.)		2,526
Net output[b]		11,192

[a] Mainly advertising, market research, commercial insurance premiums and royalties. These data were collected on a basis that differs from the rest of the census and are therefore not strictly comparable with the other data.
[b] Adjusted from the census definition by adding in expenditures by the larger firms for the purchase of transport services.
Source: Board of Trade, 1969; Edwards, 1970a.

ture on transport by manufacturing industry was about £627 million, or some 25 per cent of the £2,526 million of net output that is left when the deductions shown in Table 8.1 have been made. Clearly, even quite small variations in transport costs could have a significant effect upon the level of profits. Equally clearly, variations in any other item of expenditure will have similar effects. However, the characteristic that is thought to distinguish transport outlays from other industrial expenses is that the former are systematically related to location whereas the others are not.

The conclusion to be drawn is that inquiry among firms concerning the incidence of transport costs in various locations is necessary to establish the situation as they see it and will provide guidance on what level of inducement, if any, is necessary to persuade firms to locate in particular regions. Evidence of this kind, however, is not very suitable for policy decisions regarding where in the country firms should be induced to develop and expand, because these decisions ought to be based on different criteria – essentially those of long-term welfare, of which the cost of transport is but one part.

Spatial demand and supply cones

If we follow formal location theorists in the school of Lösch (1954) and Isard (1956), we may conceive of each location having its own spatial demand cone and an equivalent supply cone. These arise from the spatial

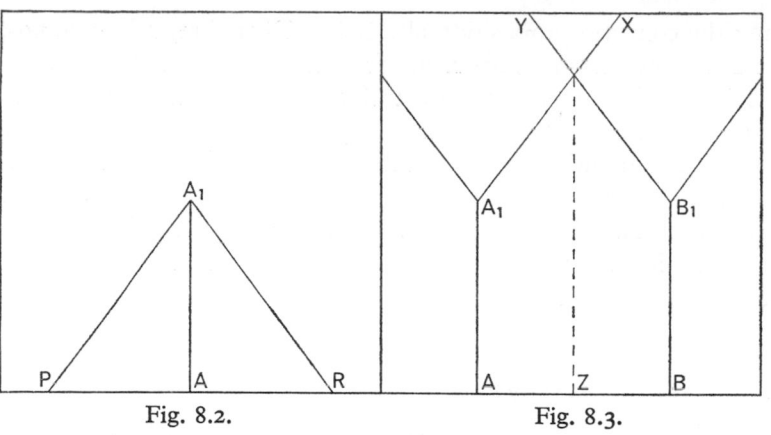

Fig. 8.2.   Fig. 8.3.

Figure 8.2. The market area for an isolated firm.
Figure 8.3. The market areas of firms in competition.

friction of moving goods combined with price elasticities of supply and demand.

A simple case is that of firm A in Figure 8.2. The height of the column represents the volume of freight leaving the premises and lines PA_1 and RA_1 show how the amount passing each point diminishes with distance from A, until all the output is sold. Rotated about its axis, the cone defines the market area for the firm, with radius $AP = AR$. Should there be a competing firm B (Figure 8.3) and should the inverted cones now represent the price delivered to the customer, there will be a point at which the customer is indifferent as to which supplier is patronised, defined as the intersection of A_1X and B_1Y projected vertically to Z. Here we have the basis for the division of the national territory into market areas. Furthermore, depending upon the slopes of the delivered price curves and other considerations, some market areas are more extensive than others and a hierarchy of market areas may be conceived.

As a result, we expect the volume of goods transported between any pair of locations to be some inverse function of the distance that separates them. This expectation accords with findings reported by many authors for a wide range of commodities, and other phenomena, in diverse countries. For some commodities, the volume of traffic falls rapidly as distance increases but in other cases the decline is much less dramatic. The exponents of these negative exponential distance–decay functions may be interpreted as a measure of the spatial elasticity of demand. As a first general proposition, the greater the distance exponent the more local the traffic and the less will inter-regional location affect the volume of freight movements; the converse proposition also holds. Therefore, if it were possible to measure the distance exponents for various commodities and also to assess the relative volumes of these traffics we could identify those the ton-mileage of which would be significantly affected by differences in inter-regional location and those for which it would not be. This would enable us to make a quantitative assessment of the transport implications of locational choices made by government. To pursue these questions, it is necessary to examine several sources of data because there is no one ideal data set to which we can turn.

National freight surveys

The Ministry of Transport has conducted two national surveys of road freight traffic, one in 1962 and another in 1967/8. Results of the former have been published (Ministry of Transport, 1964) and the origin–destination matrices by 107 zones covering all mainland Great Britain have kindly been made available by the Ministry for fifteen commodity classes covering all road freight. At the time of writing (August 1970), not even the preliminary results of the 1967/8 survey were ready. Consequently, reliance must be placed on the 1962 survey and its derivative for 1964: this is an up-dated version of the earlier survey using 78 × 78 origin–destination matrices with railway freight analysed on the same basis. The 1964 matrices have not been published but have been supplied by the Ministry on the basis of eleven commodity groups. There is also a more recent national sample survey of freight consignments by manufacturing industry (Bayliss and Edwards, 1970).

The 1962 road freight survey

In 1962, 'C' licensed vehicles (vehicles licensed to carry the goods of the owner, not for hire or reward) carried 738.5 million tons of freight and performed 16,303 million ton-miles (26,651 million ton-km), respectively 59 per cent and 49 per cent of the total road traffic. The activities of the 'C' licensed vehicles were distributed among several kinds of work, as shown on the left of Table 8.2. The work categories are fairly readily identified as 'local' business and 'regional or national' and have been assigned to these two classes. Clearly, some of the retail delivery, for example, takes place over large distances and some export deliveries may be just round the corner to the local dock. However, the distinction shows broadly that traffic which is essentially servicing local activities and is not significantly affected by inter-regional location and that which likely will be influenced

TABLE 8.2. *Great Britain: estimated work done by 'C' licensed road vehicles, analysed by type of work done, 1962*

Type of work	Local deliveries		Regional or national deliveries	
	Ton-miles, million	Mean haul, miles	Ton-miles, million	Mean haul, miles
Retail delivery	2,969	19	—	—
Maintenance and repair work	302	12	—	—
Carriage of materials to or from building sites	2,786	17	—	—
Wholesale delivery	—	—	5,157	33
Delivery of materials and fuel to factories	—	—	2,574	25
Delivery of export goods to docks	—	—	121	36
Sub-total of above	6,057	17	7,852	30
Other[a]	1,042	(19)	1,352	(19)
Total	7,099	—	9,204	—

[a] The category 'Other' has been allocated in the ratio of 6,057: 7,852 or 43.5: 56.5.
Note: all measurements are in Imperial units.
Source: Ministry of Transport (1964), *Survey...Part 1*, 52.

by macro-differences in location. The allocation of the 'Other' traffic is in proportion to the ton-miles of all the rest. It will be observed that the mean haul for the category 'Other', 19 miles (31 km) is very similar to that for the 'Local' deliveries group as a whole; therefore, it is probable that substantially more than 43.5 per cent of this 'Other' traffic ought to be regarded as local. Thus, the total of 7,099 million ton-miles (11,605 million ton-km) for local traffic substantially understates the true volume.

As a minimum figure, we may say that 7,099 million ton-miles (11,605 million ton-km.) of 'C' licensed freight was essentially local traffic unlikely to be affected by macro-location. Table 8.3 shows that this amounts to 21 per cent of all road traffic in that year, or 11 per cent of freight movements by all modes. It is unlikely that the proportion of all freight traffic that services purely local needs is as low as 11 per cent or as high as the 43.5 (probably nearer 50.0) per cent shown for 'C' licensed vehicles. However, these two proportions may be regarded as the limits within which the true figure lies.

Thirty-four commodity groups were distinguished by the 1962 freight

TABLE 8.3. *Great Britain: 'local' road freight in relation to all freight, 'C' licensed vehicles, 1962 (million ton-miles)*

	Total	'Local' road	'Local' road as % of total
All road	33,617	7,099	21
All modes	65,800	7,099	11

Note: all measurements are in Imperial units.

Sources: Ministry of Transport, 1964, *Survey... Part 1*; Central Statistical Office, *Annual Abstract of Statistics*. See Table 8.2. The figure of 65,800 differs from the figures published; an adjustment has been made to the coastal shipping estimates to provide a more realistic figure than the published one.

survey and for these classes data are available for the tonnage and ton-mileage of traffic and, by derivation, the mean haul. The full array is shown in Table 8.4 and it will be seen that the mean haul ranged from 63.8 miles (102.7 km) in the case of oil seeds, nuts etc. to 14.0 miles (22.5 km) in the case of crude minerals other than ore. If one takes account of the mean haul and knowledge of the structure of the industries that produce and use the goods, it is possible to identify nine commodity groups where, *prima facie*, inter-regional location will make virtually no difference to the mean length of road haul (Table 8.5). Coal and coke has been included because it is subject to fierce competition from petroleum, gas and electricity and consequently displays marked spatial elasticity in demand: cement, fertilisers and petroleum products are all supplied from regionally situated production/distribution points that are located in all the major regions; the other five items are essentially locally delivered in relation to intra-regional activities. This distinction between local and non-local freight is clearly not a clear-cut one, because in both categories there is a distribution of traffic about the mean length of haul.

In addition to the items shown in Table 8.5, a substantial portion of the categories 'beverages' and 'unallocable loads and mixed' represents local traffic. Together, they account for 2,824 million ton-miles (4,616 million ton-km) at a mean haul of 26 miles (42 km). However, taking a conservative view, we may say that some 12,475 million ton-miles (20,393 million ton-km) of traffic may be regarded as 'local' in the sense defined. This total is larger than the total for 'C' licensed vehicles alone (7,099 million ton-miles, see Table 8.2); 12,475 million ton-miles is equivalent to 37 per cent of all road freight, whereas 7,099 equals nearly 44 per cent of traffic carried by 'C' licensed vehicles. Since 'C' licensed vehicles have a special role in

TABLE 8.4. *Great Britain: distribution of road freight traffic by commodity and length of haul, 1962*

Commodity	Mean haul, miles	Ton-miles, million	Commodity	Mean haul, miles	Ton-miles, million
Oil seeds, nuts and kernels, animal and vegetable oils and fats	63.8	134	Animal feeding stuffs	31.7	913
			Cement	31.7	492
Chemicals and plastic materials	55.1	1,421	Crude and manufactured fertilisers	31.1	311
Iron and steel, finished and semi-finished products	52.0	2,074	Laundry and dry cleaning	30.9	99
Electrical and non-electrical machinery; transport equipment	49.9	2,201	Petroleum and petroleum products: gas	30.5	1,688
			Beverages	29.7	983
Non-ferrous metals	47.6	466	Non-ferrous metal ores	28.8	147
Flour	42.9	330	Wood, timber, cork	28.2	700
Miscellaneous manufactured articles	42.8	2,307	'Other' crude materials	26.5	467
Fish	41.1	156	Lime	26.4	169
Textile fibres and waste	38.8	291	Building materials	25.7	2,785
Empty containers	37.3	590	Tars from coal and natural gas	25.3	139
Metal manufactures	36.6	764	Dairy produce, eggs	24.0	1,167
Meat and poultry	35.2	433			
Furniture removals	34.5	100	Iron ore and scrap iron	23.1	233
'Other' foods, tobacco	33.6	2,456	Unallocable loads, mixed loads	21.7	1,841
Fresh fruits, vegetables, nuts and flowers	33.4	1,149	Coal and coke	14.2	2,134
Cereals	32.9	517	Crude minerals other than ore	14.0	3,630
Live animals	32.4	330	**Total**	**26.9**	**33,617**

Note: mean haul has been calculated to one decimal place only for purposes of ranking. All measurements are in Imperial units.

Source: Ministry of Transport, 1964, *Commodity Analysis*, 23.

TABLE 8.5. *Great Britain: road freight traffic,*
nine 'local' commodities, 1962

	Ton-miles, million	Mean haul, miles
Cement	492	32
Crude and manufactured fertilisers	311	31
Petroleum and petroleum products; gas	1,688	31
Coal and coke	2,134	14
Laundry and dry cleaning	99	31
Lime	169	26
Building materials	2,785	26
Dairy produce	1,167	24
Crude minerals other than ore	3,630	14
Total of the nine commodities	12,475	19
Total, all commodities	33,617	27

Note: all measurements are in Imperial units.
Source: Ministry of Transport, 1964, *Commodity Analysis*, 23; see also Table 8.4.

local traffic, these two sets of data are consistent with each other. A total of 12,475 million ton-miles represents 19 per cent of freight traffic by all modes in the year 1962.

Thus far, we may conclude that about two-fifths of all road freight may be regarded as invariant with inter-regional location. For the present, we will assume that all the traffic moving by rail, coastal shipping and inland waterways is of a character that will be affected by macro-location. On this basis about one-fifth of freight transport is serving entirely local needs.

The 1964 freight survey data

O'Sullivan (1971) has conducted some experiments with interaction models, using the eleven commodity groups that are available in 78×78 origin–destination matrices covering the whole of Great Britain. Using a gravity model formulation, he obtained the distance exponents (or β coefficients) that are shown in Table 8.6. These may be interpreted as a measure of the spatial elasticity of demand for the commodities in the context of the given distribution of population and production and the actual flows of traffic that are observed. A low coefficient is equivalent to a low elasticity and the converse. A striking feature of this table is the close association between the magnitude of the coefficient and the degree of explanation achieved by the gravity model; a Spearman rank correlation gives an R value of 0.93,

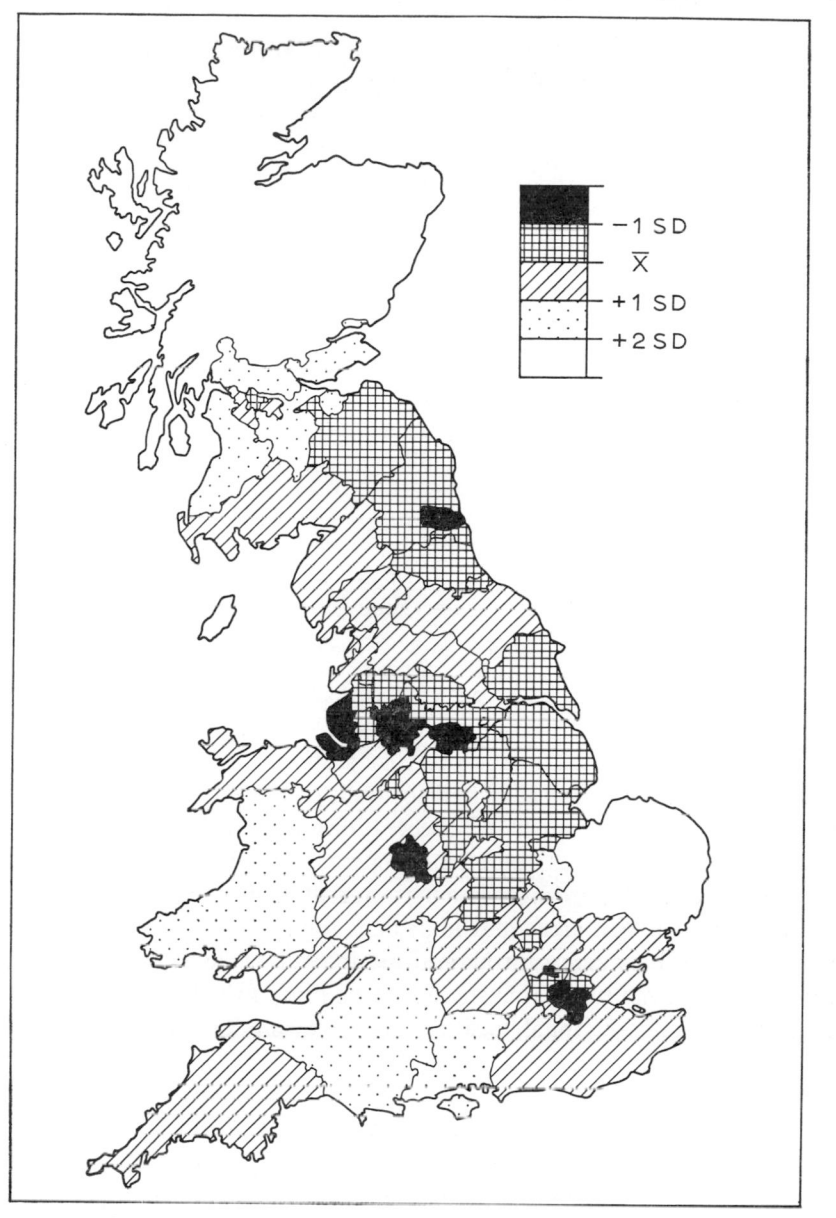

Figure 8.4. Great Britain: road freight β values mapped as standard deviations about their mean, 1964.

Source: O'Sullivan, 1971.

TABLE 8.6. *Great Britain: gravity model distance exponents and freight tonnage, road traffic by commodity, 1964*

Commodity group	β	R^2	Tonnage of freight	
			Million long tons	%
Steel	−0.77	0.24	45.5	3.3
Chemicals	−0.80	0.31	34.9	2.5
Scrap	−0.82	0.28	11.2	0.8
Transport and equipment	−0.96	0.41	8.0	0.6
'Other' manufactures	−1.09	0.43	150.3	10.9
'Other' crude materials	−1.16	0.45	69.4	5.0
Oil	−1.17	0.37	57.8	4.2
Miscellaneous	−1.36	0.52	113.2	8.2
Coal	−1.46	0.49	142.1	10.3
Foodstuffs	−1.53	0.62	306.8	22.2
Building materials	−1.64	0.60	444.4	32.0
Total			1,383.6	100.0

Source: O'Sullivan, 1971.

which is acceptable at the 99 per cent probability level. Furthermore, the commodity groups with large exponents are clearly comparable with the 'local' traffics discussed in the previous section.

Exponents cannot be calculated for each commodity for each of the 78 zones because the number of observations becomes too small to have meaning. O'Sullivan has obtained exponents for each zone on the basis of aggregate traffic (Figure 8.4). The pattern that emerges is of low exponents for the major urban areas and high ones for the rural areas, rather than a distinction between central and peripheral regions of the country, though O'Sullivan noted that there is some hint of the latter. The implication is that intra-regional location affects mean haul more than inter-regional location does.

Distribution of traffic by distance zones, 1962 and 1964

The published data for the 1962 road freight survey allocate freight into distance bands, as shown in Table 8.7. No breakdown by commodities is given. If we supposed that all freight moving less than 25 miles (40 km) represents 'local' traffic in the sense that we have been using the term, then 21 per cent of the ton-miles is to be so classed. Now this is a crude measure of 'local' traffic and it would clearly be desirable to have a finer breakdown

TABLE 8.7. *Great Britain: distribution of road freight by distance bands, 1962*

Distance	Tons of freight, million	Ton-miles of freight, million	Mean haul, miles	% of ton-miles
Under 25 miles	821.8	7,041	8.6	21
25–49 miles	197.1	5,945	30.2	18
50–99 miles	134.0	7,454	55.6	22
Over 99 miles	95.2	13,177	138.4	39
Total	1,248.1	33,617	26.9	100

Note: all measurements are in Imperial units.
Source: Ministry of Transport, 1964, *Survey...Part 1*, 45–8.

TABLE 8.8. *Great Britain: estimated distribution of road freight traffic by ten-mile distance zones, 1964*

Distance zones, miles	Estimates from the 78 × 78 matrix		Adjusted estimates	
	Tons of freight, million	Ton-miles of freight, million	Ton-miles adjusted to 1964 total	% of adjusted ton-miles
1–10	381.9	1,909	1,650	4.2
10–20	365.0	5,457	4,750	12.2
20–30	271.5	6,538	5,650	14.5
Total	1,018.4	13,922	12,050	30.9
Over 30	365.2	31,132	26,950	69.1
Grand total	1,383.6	45,054	39,000	100.0

Note: all measurements are in Imperial units.
See text.

of distance zones. Such a breakdown can be obtained from the origin–destination matrices for either 1962 or 1964; Table 8.8 presents some estimates for 1964.

Each flow recorded in the 78 × 78 origin–destination matrix for road traffic has been assigned to a distance interval ranging from 1 to 10 miles, 11 to 20 miles, etc. at ten-mile (16 km) intervals up to 410 miles (660 km). This gives the frequency distribution of the tonnage of freight moving over various distances. The distances so assigned are based on road distances between the centroids of the traffic zones and on an estimate of the length of intra-zonal hauls (at two-fifths the square root of the area): they are therefore subject to considerable error. If we assume that for each distance zone, the mean haul is half way between the upper and lower limits (5, 15, 25, etc. miles), an estimate can be made of the ton-miles. The total so obtained, 45,054 million ton-miles (73,650 ton-km), exceeds the estimated national volume for that year by 6,000 million ton-miles. This is due to the skew distribution of journeys, such that the mean haul in each distance zone is less than the assumed mid-point value. Making a proportionate adjustment yields the third column of Table 8.8. Comparison of this table with Table 8.7 shows that there is a reasonable coincidence in the distributions. Thus, although the data in Table 8.8 are not very reliable, there seems to be no doubt that very short hauls of under 10 miles (16 km) are unimportant in terms of ton-miles despite accounting for nearly one-third of the tonnage. Short haul traffic is concentrated in the distance bracket of 11–30 miles (18–48 km).

A sample consignment survey, 1966–7

Bayliss and Edwards (1970) have reported on a national sample survey of manufacturing firms, in which data were gathered for individual consignments. Several points emerge from this survey that are germane to the present argument. The first is summarised in Table 8.9, showing the distribution of consignments by commodity and distance zones. The commodity groups do not exactly match those previously quoted but it will be observed that there is a general correspondence in the ranking shown in Table 8.6 and for the sample survey data (Table 8.9), which are ordered by the proportion of freight moving under 25 miles (40 km). Particularly striking is the very large share of consignments of foodstuffs and building materials moving less than 25 miles (40 km) or even 50 miles (80 km). Together, these two commodities accounted for 20,070 consignments or 32 per cent of the total. This particular point is amplified in the text:

The foodstuffs group of commodities covers both raw and processed foods, but in the present survey, since only outwards consignments from food processing industries have been covered, most of these commodities consist of processed foods. Also, for a better understanding of the nature of the consignments of this industry it should be noted that a high proportion of the movements would probably be from processing plants direct to retail outlet. Some inter-plant movements obviously occur because, for example, the commodity group includes flour which would move mainly from the mill to bread or biscuit factories, and there are also movements from processing plants to regional warehouses or wholesale distribution

TABLE 8.9. *Great Britain: distribution of consignments by length of haul, all modes, 1966–7*

	Cumulative % of consignments						Number of consignments in survey sample	
	Under 25 miles	Under 50 miles	Under 100 miles	Under 150 miles	Under 200 miles	200 miles and over	Number	%
Chemicals	6	26	48	59	79	100	4,060	6.4
Engineering and electrical goods	8	31	49	75	81	100	3,020	4.7
Non-ferrous metals	13	18	27	46	63	100	930	1.5
Metal manufactures	15	26	39	52	68	100	2,970	4.7
Iron and steel	17	28	53	72	85	100	1,720	2.7
Transport equipment	23	37	55	71	85	100	2,620	4.1
Crude materials	25	32	56	73	86	100	2,050	3.2
'Other' manufactures	32	43	59	74	85	100	26,330	41.3
Foodstuffs	57	80	86	92	95	100	16,410	25.7
Building materials	59	77	84	87	91	100	3,660	5.7
Total	35	50	64	77	87	100	63,770	100.0

Note: commodities ranked by % of consignments sent under 25 miles.
Source: Bayliss and Edwards, 1970, 46, 74.

centres, but in the main the distribution structure of many parts of the food and drink industry is such that many consignments probably flow direct to the shops or public houses (Bayliss and Edwards, 1970, 29–30).

The building materials industry, like the foodstuffs industry, produces mainly for points of final consumption (in this case the construction industry) (Bayliss and Edwards, 1970, 34).

Conclusion from the freight traffic data

Given the nature of the data that are available, the analysis so far has been crude. Nevertheless, we may be confident that at least one-fifth of road freight may be regarded as 'local' and unaffected by inter-regional location. The true proportion is probably about double this, at something like two-fifths. If we assume that none of the rail and shipping cargoes can be classed as 'local' in the sense used here, this still means that about one-fifth of all freight movements are 'local'. The question now arises whether this order of magnitude tallies with other evidence, notably that available from the 1963 census of production industries?

The 1963 census of production

The 1963 census of production industries obtained data on the expenditure by firms for the purchase of transport services and, in the case of the larger firms, their outlay on own-account transport. Since the majority of consignments are quoted at delivered prices, a relatively small outlay is incurred by firms for the inward movement of consignments. Thus, the census data relate almost exclusively to the cost of transport from establishments. It is important to bear this in mind, since the recorded transport costs for each industry do not give a true picture of its total transport demands. However, without indulging in double-counting, there is no way round this problem.

At the time of writing, the regional tables had not been published and it is therefore necessary to work with the national aggregate data. As reported by the Board of Trade (1969) for the larger firms, total expenditure on transport was £722.9 million, of which manufacturing industry accounted for £568.6 million and the non-manufacturing industries £154.3 million, or 21 per cent of the total. The non-manufacturing industries comprise Orders II, XVII and XVIII – mining and quarrying, construction, gas, electricity and water. As Table 8.10 shows, these three industries are characterised by exceptionally short average hauls and may therefore be fairly described as serving local markets. Even ignoring mining and quarrying, which includes coal, the other two non-manufacturing industries spent £113.7 million, or 16 per cent of the total expenditure.

Edwards (1970a) has adjusted the 1963 census data in a number of ways and obtained a substantially higher total outlay on transport than the census shows; his figure is £943.1 million. Of this total, £627.0 million arose from the manufacturing industries, and the non-manufacturing sector accounted for the remaining £316.1 million, which is 30 per cent of the aggregate. Ignoring Order II, mining and quarrying, the other two non-manufacturing industries on Edwards' reckoning spent £148.3 million, which is equivalent to 16 per cent of the grand total. The much greater importance attaching to Order II in the Edwards' calculations than in the census returns arises from his inclusion of merchants' transport costs on the collection and delivery of coal, which is usually sold at a pit-head price and therefore no cost of transport on delivery is returned by the coal industry for census purposes. We may conclude that a minimum of about 21 per cent of transport outlays in the production industries is concerned with essentially local traffic generated by the non-manufacturing sector. On Edwards' basis of reckoning, the proportion is nearly one-third.

TABLE 8.10. *United Kingdom: selected data for production industries, 1963*

	Transport as % of net output[a]	Mean haul, 'C' licensed vehicles, miles[b]	Total transport costs, £ million, census data, large firms[c]	Total transport costs, £ million, all firms (Edwards)[d]
VII. Shipbuilding and marine engineering	1.1	32.2	2.4	2.6
VIII. Vehicles	2.0	50.0	23.5	25.0
XII. Clothing and footwear	2.05	40.0	7.9	9.3
VI. Engineering and electrical goods	2.4	42.0	61.1	67.0
X. Textiles	3.0	23.2	24.8	26.9
XI. Leather, leather goods and fur	3.3	5.9	2.0	2.5
XVI. 'Other' manufacturing industries	3.9	40.0	15.8	17.5
IX. Metal goods n.e.s.	4.1	47.8	26.1	30.8
XV. Paper, printing and publishing	5.1	30.3	44.6	48.5
V. Metal manufacture	6.1	35.3	53.0	55.1
IV. Chemicals and allied industries	6.3	33.1	69.6	74.3
XIV. Timber, furniture, etc.	6.6	40.9	20.3	26.2
III. Food, drink and tobacco	11.9	32.1	159.3	176.9
XIII. Bricks, pottery, glass, cement, etc.	12.6	22.8	58.2	64.4
Total	5.1	25.0	568.6	627.0
XVIII. Gas, electricity and water	2.5	11.1	40.6	26.9
XVII. Construction	4.8	13.3	88.8	121.4
II. Mining and quarrying	5.1	12.9	24.9	167.8
Total	4.3	13.0	154.3	316.1
Grand total	4.9	22.1	722.9	943.1

[a] Transport costs of larger firms only. Net output adjusted by adding in expenditures of larger firms on the purchase of transport services. Board of Trade, 1969.
[b] 1962 road freight survey figures. Ministry of Transport, 1964, *Survey...Part 1.*
[c] As recorded by the census for the larger firms. Board of Trade, 1969.
[d] Estimates made by Edwards, 1970a.

Interest therefore turns to the manufacturing sector, Orders III–XVI inclusive. Table 8.10 sets out some summary data for the main SIC industrial classes, which are ranked according to the proportion transport cost is of net output. The second column records the mean haul on freight moved by 'C' licensed vehicles in 1962. As these vehicles handle only a variable proportion of the total traffic generated by each industry, the mean-haul data must be taken as indicative only. However, it is generally true that the shorter the mean haul the greater the proportion of traffic moved by the 'C' licensed fleet and therefore the more reliable the figure for the average haul. The third and fourth columns show the cost of transport for the larger firms as reported in the census and Edwards' estimates of the total for all firms.

There is some suggestion that those manufacturing industries for which transport costs are low in relation to net output have a longer average haul than do the others. Using a Spearman's rank correlation, however, the relationship is only just significant at the 95 per cent level even when the leather and ship-building industries are eliminated. The leather industry is quite clearly subject to unusual conditions with a mean haul of only 6 miles (10 km); the shipbuilding industry is also exceptional because most of its output sails from the shipyards. With these data, at least, the hypothesis cannot be substained that long hauls are associated with low relative transport costs though such an association is to be expected on *a priori* grounds.

We may regard Orders III and XIII as industries that primarily supply the local market and for which inter-regional location has relatively little effect. Indeed, the strikingly high proportion of transport costs to net output in these two industries confirms their sensitivity to the cost of moving freight. Taken together, the larger firms in these two industry classes spent £217.6 million on transport in 1963, amounting to 38 per cent of the total recorded for manufacturing industry or 30 per cent of all production industries.

If we take the non-manufacturing industries plus Orders III and XIII, the proportion of transport expenditure as reported in the census amounts to 51 per cent. On the basis of Edwards' adjusted figures, the proportion is almost identical at 53 per cent, comprising 30 per cent for the non-manufacturing industries and 23 per cent for Orders III and XIII. These census data cover expenditures on all modes of transport and yet the proportion of outlays attributable to 'local' traffic is somewhat higher than the proportion of ton-miles shown for the operation of 'C' licensed road vehicles (see p. 220).

The remaining industries may be regarded as ones where the volume of freight traffic is affected by macro-location. In making this dichotomous division, some violence is undoubtedly done to the data but we may reasonably assume that the errors approximately cancel and that the resulting order of magnitude bears some relation to reality.

The census data just considered take no account of the wholesale and retail demands for transport. However, this consideration does not apply to the non-manufacturing Orders XVII and XVIII, and in the Edwards' data allowance has been made at least for the wholesaling element of transport demand arising from the coal industry. The food and beverage industries are characterised by a high proportion of sales direct to retail outlets and similarly bricks in particular are generally delivered to the construction site. Indeed, large-scale movement to wholesale outlets is generally common only for those industries which sell on a national market.

Conclusion

Evidence from the 1963 census of production industries suggests that up to one half of the expenditure on transport is by industries that serve essentially local markets. Consequently, inter-regional location will have a small effect on the volume of freight and total transport outlays for this portion of freight movement. These census data indicate a substantially higher proportion of 'local' traffic than is evident from the traffic data, a discrepancy that may well arise from the high level of aggregation used. It is difficult to carry the analysis to the scale of the Minimum List Heading owing to the lack of independent traffic data at this level of industrial disaggregation. However, another reason why expenditures on 'local' traffic may be substantially higher than the volume of traffic would appear to warrant is the structure of freight charges: short hauls may well be more costly per ton-mile than long hauls are. This question is examined in the next section.

Distance and the cost of transport

The greater the distance a consignment is sent, the greater is the cost of shipment if all other things are equal. At the simplest, one may assume with Weber that costs are directly proportional to distance, or that some degree of tapering occurs. But main interest attaches to the relationship between terminal costs and movement costs, since the larger the former are relative to the latter, the less is the significance of increments in distance. Our problem is to measure the relationship between terminal charges and movement charges. Most of the data in this field are based on a rather different concept, namely, the distinction between vehicle standing costs and running costs; this is clearly important from the operators' point of view and is valid in accountancy terms but is not very closely related to the distinction between terminal costs and movement costs. For our purpose, the requirement is to obtain data on actual freight charges in relation to distance.

Bayliss and Edwards (1970) obtained a national sample of freight consignments made by industries and their data include information on the charges for consignments sent by rail and professional road haulier. They were able to 'explain' a very high proportion of the variation in charges by factors such as length of haul, the weight of consignment and the volume

TABLE 8.11. *West Cumberland Survey: average cost of consignments by public transport, 1966*

Weight category of consignment	Cost per ton, £	Cost per ton-mile, new pence
up to 22 lb	10.246	90.4
22–56 lb	3.929	28.7
56–112 lb	2.933	20.0
112–560 lb	1.504	15.0
560–2240 lb	0.679	5.0
1–2 tons	0.750	4.6
2–3 tons	0.521	3.3
3–4 tons	0.383	2.9
4–5 tons	0.250	2.1
5–7 tons	0.250	2.1
7–10 tons	0.179	2.1
10–15 tons	0.154	1.7
15–20 tons	0.217	1.7
20 tons and over	0.212	1.7

Note: all measurements are in Imperial units.
Source: Edwards, 1967, *Statistical Appendix*, 26.

of freight annually shipped from the establishment by the mode (road or rail) in question. Estimating equations were obtained for the two modes separately and 'in both models consignment weight was by far the most significant variable, explaining more than four-fifths of the variation in charges' (Bayliss and Edwards, 1970, 159). Length of haul proved to be comparatively unimportant as an independent factor. Data from the West Cumberland Survey provides dramatic support for the proposition that consignment size is an important factor affecting unit charges, as is shown in Table 8.11.

This interpretation cannot be taken at face value, since the size of consignment is correlated with length of haul and also with the volume of traffic. This is clearly demonstrated by evidence from the Severnside Study (Edwards, 1970b). Using Spearman's rank correlation for consignments of textiles, fabrics and yarns, clothes and hosiery, an r_s value of 0.83, significant at the 99.8 per cent level, was obtained between volume of flow and consignment size. Consequently, the Bayliss and Edwards results are not very helpful in the present context, though they do suggest that it is the size of consignment, its nature and the volume of traffic that are the main determinants of cost, rather than distance as such. However, if there is a high correlation between these factors and distance, then the distance variable is a useful one to use.

Deakin and Seward (1969) also carried out a national sample survey, this time of hauliers. Road haulage firms were asked the average charge they made and the mean distance of haul for each of 34 commodity groups. The unpublished summary data have kindly been made available to the author. For some commodities there were fewer than 15 usable observations and in a few cases where data were more numerous no meaningful linear regressions could be computed. Nevertheless, for 23 commodities regressions of the form

$$Y = a + bX,$$

where Y equals transport charge in shillings per ton and X equals length of haul in miles yielded results acceptable at the 99 per cent level. The results are set out in Table 8.12.

The Deakin and Seward data were obtained from a sample of firms and there is considerable variation in the volume of traffic handled by them; no weighting has been used in calculating the regressions. Some of the observations appeared to be clearly maverick, being vastly too high or low a charge for the given distance or being an exceptional length of haul; these data have been excluded from the calculations. Table 8.12 shows the number of eliminated observations and the number (N) used for the regressions. Consequently, the coefficients shown in this table must be treated with care and regarded as giving no more than orders of magnitude for the listed commodities.

The intercept of the regressions (a in Table 8.12) may be interpreted as the terminal charge which is invariate with distance; the slope, or b coefficient, indicates the increment in cost in shillings per ton for each mile of additional haul. Dividing a by b shows the distance a consignment must be sent to incur a movement cost equal to the terminal cost. The unweighted average coefficients and the median values for the 23 commodities may be compared:

	a	b	a/b
Unweighted mean	15.561	0.197	79
Median	15.817	0.200	79

The structure of road freight charges is such that a transit distance of almost 80 miles (129 km) is required to double the charge made by a haulage contractor. This figure may be compared with evidence from two other countries. Harris (1954) compiled figures for both road and rail traffic in the USA for the early nineteen-fifties; these show that the cost of haulage

TABLE 8.12. *Great Britain: road haulage charges in relation to length of haul, 1966*

Commodity	Observations Unused	N	a	b	a/b	R²
Electrical and non-electrical machinery and transport equipment	0	29	27.736	0.161	172	0.446
'Other' chemicals and plastics	2	32	21.658	0.187	116	0.612
Metal manufactures	5	28	21.186	0.185	115	0.687
Iron and steel, finished and semi-finished	2	48	21.100	0.147	144	0.609
Wood, timber and cork	2	37	19.560	0.155	126	0.629
Mixed loads	3	34	19.499	0.200	97	0.644
'Other' manufactured goods	2	36	19.133	0.209	92	0.548
'Other' foods and tobacco	7	29	17.888	0.216	83	0.638
Petroleum and petroleum products: gas	0	16	17.676	0.201	88	0.671
Flour	0	18	16.474	0.148	111	0.866
Fresh fruit, vegetables, nuts and flowers	7	19	16.259	0.241	67	0.801
Textiles (fibres and waste)	1	15	15.817	0.269	59	0.879
Oilseeds, etc.	1	26	15.354	0.190	81	0.818
Beverages	2	17	14.420	0.247	58	0.913
Fertilisers	0	44	13.968	0.144	97	0.810
Iron ore and scrap iron	3	29	13.479	0.125	108	0.634
Animal feeding stuffs	4	46	13.037	0.219	60	0.663
Cement	0	32	12.543	0.142	88	0.750
Lime	0	20	11.743	0.185	63	0.727
Cereals	4	32	10.434	0.231	45	0.815
Coal and coke	4	40	8.973	0.223	40	0.627
Building materials	10	64	7.086	0.251	28	0.730
Crude minerals other than ore	1	48	2.891	0.259	11	0.768

All the regressions are acceptable at the 99 per cent level. See p. 235.
Source: Deakin and Seward, 1970.

(not the rate charged) doubled at a distance of 200 miles for road haulage and 440 miles for rail traffic (respectively 320 km and 710 km). The greater distance required in the United States than in Great Britain for movement costs (charges) to equal terminal costs (charges) probably reflects the much more dispersed nature of the space-economy in the former country than in the latter. However, Australia is a country with an even more dispersed spatial structure than the United States but has a structure of freights that confirms the general picture for America and Britain. Rimmer (1970) collected data for both road and rail shipments in 1968 that show a linear relation with distance. The shortest haul for which he published data is 192 miles (309 km) in the case of road transport and 203 miles (327 km) for rail freight – in both cases these are distances considerably in excess of the norm for Great Britain. Rimmer's data indicate that movement charges are equal to terminal charges at about 1,000 miles (1,600 km) for both road and rail traffic, for general freight measured at 40 cubic feet per ton (1.1 cubic metres per metric ton). This brief comparison with evidence from two other countries indicates that the British data do not give an implausible ratio between terminal and movement charges.

If we possessed reliable data of the kind shown in Table 8.12 for all commodity groups and, by the same commodity classes, information on the distribution of freight by distance zones, an estimate could be made of the proportion of freight costs that is invariable with length of haul. Since linear regressions in natural numbers have been fitted to the freight charge data, it is necessary only to know the tonnage and mean haul for each commodity group. Unfortunately, we do not possess data for both freight charges and freight volume covering all the commodity classes. Therefore, two exercises only can be undertaken. The first is to apply average regression coefficients, such as the mean or median values for a and b shown above, to the aggregate of all road freight as set out in Table 8.7. The second is to take those commodities in Table 8.4 for which freight-charge data are available in Table 8.12 and apply the respective a and b coefficients to the commodity tonnages and mean hauls. The results of these exercises are set out in Table 8.13. The total road transport charge shown in this table is to be treated as notional: it is the distribution between terminal and movement charges that is important. Edwards (1969 and 1970a) estimated annual total expenditure by the production industries and wholesale trades in 1963 and 1965 respectively as £943.1 million and £404.0 million, or £1,347.1 million altogether by all modes.

The striking thing about Table 8.13 is the extraordinarily high proportion of road freight charges that may be regarded as terminal charges and therefore not varying with the length of haul – something approaching three-quarters of the total. This estimate applies to all road freight and therefore includes that element which previously has been regarded as 'local' traffic, characterised by the short length of haul. Therefore, for the non-local freight, terminal charges will be a lower proportion of total charges than for all freight, perhaps 50 per cent. In that case, inter-regional differences in location will not have a dramatic effect upon the nation's freight transport bill.

TABLE 8.13. *Great Britain: estimates of the relationship between terminal charges and movement charges for road freight (1962 freight volumes and 1966 charges)*

	Estimates of charges, £ million			Terminal charge as % of total
	Terminal	Movement	Total	
All road freight (Table 8.7) with median coefficients of $a = 15.8$ and $b = 0.200$	985.9	335.7	1,321.6	74.6
23 commodity classes of road freight (Table 8.4) (a and b coefficients from Table 8.12)	689.7	309.1	998.8	69.1

Sources: see Tables 8.4, 8.7 and 8.12.

Mobility of firms, transport costs and regional policy

Government has for many years operated a variety of policies with the intention of controlling regional development (McCrone, 1969). An important element in policy implementation has been the control of industrial and, more recently, office employment through the administration of the Industrial Development Certificates and Office Development Permits. The certificates are required for any industrial development that exceeds a critical threshold floor area; they are required both for expansion *in situ* and for firms wishing to establish entirely new plants. IDC controls do not affect small industrial plants with an employment below approximately 25 workers. There is clear evidence that the distance over which mobile firms actually move is positively related to the size of plant involved: the larger it is, the greater the average distance of move. Consequently, it is interesting to know the distribution of employment and the number of establishments according to the incidence of transport costs, and especially for plants with 25 workers or over.

In Table 8.10 the Main Order industries are ranked according to the proportion of transport costs in their net output. Using that ranking, but ignoring the non-manufacturing industries, Table 8.14 has been compiled to show the cumulative number of employees and plants. Read in conjunction with Table 8.15, it is striking just how large is the proportion of the manufacturing sector in which transport costs are relatively low: over half the larger establishments are in industries where transport costs are less

TABLE 8.14. *Great Britain: the number of establishments and number employed in manufacturing industries, cumulated by the incidence of transport costs, 1963 (Main Order classes)*

Transport as % of net output	All establishments		Establishments employing 25 workers or over	
	Number	Employment (thousands)	Number	Employment (thousands)
1.9 and under	1,281	202.6	505	195.1
2.9 and under	28,551	3,411.2	12,937	3,248.7
3.9 and under	41,354	4,524.2	19,616	4,296.2
4.9 and under	51,720	5,031.9	22,893	4,737.3
5.9 and under	61,991	5,615.8	26,539	5,261.1
6.9 and under	77,814	6,894.6	32,522	6,445.8
11.9 and under	84,849	7,650.7	35,978	7,173.4
12.9 and under	89,949	7,959.7	37,920	7,452.6
Total	89,949	7,959.7	37,920	7,452.6

Source: Board of Trade, 1969.

than 4 per cent of net output. Equally striking is the large average size of establishments in those industries where transport costs are relatively low, and vice versa. This relationship is remarkable up to the 5.9 per cent level. Where transport costs are between 6.0 and 11.9 per cent of net output, the size of plant increases again. This increase is attributable to the substantial economies of scale in manufacture in steel production and some of the food and drink industries.

The larger plants are generally more mobile over greater distances than are smaller ones (see Chapter 2): they are also the ones that are most completely controlled by IDC regulations. In this sense, therefore, they are an important 'leading' element in government regional policy. Yet it is precisely for these industries that the average incidence of transport costs is low but that the spatial variation in transport costs is apt to be considerable. The implication is that for the larger plants inter-regional location choices will make little difference to the nation's transport bill but may in some cases have an impact on the operating costs of the firm. Overall, it seems most unlikely that there is any serious divergence between private and social considerations.

TABLE 8.15. *Great Britain: the number of establishments, number employed and mean employment in manufacturing industries according to the incidence of transport costs, 1963 (Main Order classes)*

Transport as % of net output	All establishments			Establishments employing 25 workers and over		
	Number	Employment (thousands)	Average employment per establishment	Number	Employment (thousands)	Average employment per establishment
1.9 and under	1,281	202.6	158	505	195.1	386
2.0–2.9	27,270	3,208.6	118	12,431	3,053.6	246
3.0–3.9	12,803	1,113.0	87	6,680	1,047.5	157
4.0–4.9	10,366	507.7	49	3,277	441.1	135
5.0–5.9	10,271	583.9	57	3,646	523.8	144
6.0–6.9	15,823	1,278.8	81	5,983	1,184.7	198
11.0–11.9	7,035	756.1	108	3,456	727.6	211
12.0–12.9	5,100	309.0	61	1,942	279.2	144
Total	89,949	7,957.7	88	37,920	7,452.6	197

Source: Board of Trade, 1969.

Postscript on personal expenditure on transport

Edwards (1969 and 1970a), using the 1963 census of production and the 1965 census of the wholesale trades, has estimated total expenditure on goods transport as:

1963, production industries	£943.1 million
1965, wholesale trades	£404.0 million
	£1,347.1 million

These figures compare with an estimated spending of £2,297 million on passenger transport by rail, fare-paying road services and private motoring in 1966 (Deakin and Seward, 1969, 66). Private motoring absorbed over three-quarters of the money spent on personal mobility. It is clear that the greater part of transport costs are associated with the movement of people and not of freight.

The only source of information known to the author which examines the spatial variation in personal expenditure on transport is the annual Family Expenditure Survey. Table 8.16 shows the standard regions arranged

TABLE 8.16. *United Kingdom: proportion of family expenditure on transport and vehicles, 1965–7*

Standard region	Average weekly expenditure, pounds per person		Expenditure on transport and vehicles as % of total expenditure		Rank difference
	£	Rank	%	Rank	
GLC area of South East	9.154	1	13.27	3	(−2)
Rest of South East	7.860	2	13.57	2	(0)
West Midland	7.562	3	12.72	4	(−1)
North West	7.271	4	11.34	8.5	(−4.5)
East Anglia	7.181	5	14.80	1	(+4)
South West	6.956	6	11.55	6	(0)
Yorkshire and Humberside	6.954	7	10.37	12	(−5)
Scotland	6.942	8	11.05	11	(−3)
Wales	6.937	9	11.15	10	(−1)
East Midlands	6.860	10	11.60	5	(+5)
North	6.629	11	11.39	7	(+4)
Northern Ireland	5.717	12	11.34	8.5	(+3.5)
United Kingdom	7.442	—	12.19	—	—

Source: Family Expenditure Survey as published in *Abstract of Regional Statistics*, 5, 1969, 73 (Central Statistical Office, 1969).

according to the average weekly spending per member of a family and the associated proportion spent on 'transport and vehicles'. A Spearman rank correlation gives a value of r_s of 0.59, which is acceptable at the 95 per cent level of probability but not at the 99 per cent level. The divergences in rank ordering are as interesting as the general correspondence and are shown in the table. The pluses indicate regions in which transport outlays are higher than might be expected on the basis of average expenditure; minuses show the reverse position. Five regions have differences in the range ±2 and they include the South West and the Greater London Council. The three regions with a negative difference of 3 or greater are all primarily urban and industrial, whereas two of the remaining group, where transport outlays are relatively large, have a high proportion of their inhabitants in rural and small-town settlements.

There is a strong suggestion of an income-elasticity of demand for transport, which is an interpretation consistent with international evidence on income-elasticities of demand. A further factor of importance is the intra-regional distribution of population; the more dispersed the distribution, the higher travel costs are likely to be. However, there is no clear indication that location within the country – the inter-regional effect – is of particular importance.

In the absence of better evidence, we may conclude that personal expenditure on transport is not significantly affected by location of residence at the inter-regional scale. This conclusion is not especially well based on statistical evidence but appears to be the conclusion to be drawn from the available data.

Conclusion

Using data of the kind that have been presented in this chapter lays one open to the strong possibility that the argument is circular in nature. In particular, it may be the case that the relatively high proportion of freight movements that may be described as 'local' in character and therefore likely not to be affected by inter-regional location arises from the advantages of spatial agglomeration and the intense local interaction that can be fostered in the major conurbations. If this interpretation is in fact correct, then much of the argument here presented is empty. However, given the available evidence on the nature of the short-haul traffic and especially the commodities involved, it is clear that local movements are dominated by commodities that are not of a character normally associated with regional external economies of scale. Thus, although there must be some element of circularity in the argument, it is unlikely to be sufficient to vitiate the main conclusions.

It seems likely that at least one-fifth of the freight volume moving in this country is related to industries that are closely tied to their markets. For this proportion of freight, inter-regional location of population increments in urban areas of at least half a million people will imply virtually no change in the total demand on transport resources. In terms of expenditures on transport by classes of industry, the proportion of outlays associated with the market orientated activities approaches one half. A major reason for the proportion of expenditures on local traffic being higher than the volume thereof derives from the structure of freight rates and the great importance of terminal costs relative to movement costs – something approaching three-quarters in the road haulage sector. The implication is that inter-regional location makes very little difference to the total demands made upon the transport sector and consequently the consumption of real resources.

It must be remembered that a limited concept of transport cost has been used in this discussion, that the data do not cover the total cost of moving freight and the associated costs of packaging, etc., and that no account has been taken of public infrastructure expenses. Nevertheless, the evidence here presented strongly suggests that inter-regional location makes little difference to the nation's freight transport bill; such a conclusion implies that decisions whether to continue with attempts to disperse population from London and the Midlands into the Development Areas and to bolster the economies of the less fortunate regions should be taken on grounds other than that of the cost of moving freight.

Acknowledgement

This paper presents some of the work done under a project financed by the Social Science Research Council and co-directed by Dr P. M. O'Sullivan, Department of Geography, University of Bristol. I am indebted to Mr R. Dennis for help with the calculations.

References

Bayliss, B. T. and S. L. Edwards (1970). *Industrial Demand for Transport*, Ministry of Transport, HMSO.

Board of Trade (1969). *Report on the Census of Production 1963*, CXXXI, *Summary Tables 1–12*.

Brown, A. J. (1969). 'Surveys in applied economics: regional economics, with special reference to the United Kingdon', *Economic Journal*, LXXIX, 759–96.

Caesar, A. A. L. (1964). 'Planning and the geography of Great Britain', *Advancement of Science*, XXI, 91, 230–40.

Cameron, G. C. and G. L. Reid (1966). *Scottish Economic Planning and the Attraction of Industry*, Oliver and Boyd, University of Glasgow Social and Economic Studies, Occasional Papers No. 6.

Central Statistical Office. (Annually) *Annual Abstract of Statistics*, HMSO.

Central Statistical Office. (1969). *Abstract of Regional Statistics*, 5, HMSO.

Chisholm, M. (1970). *Geography and Economics*, Bell, 2nd rev. edition.

Clark, C. (1966). 'Industrial location and economic potential', *Lloyds Bank Review*, 82, 1–17.

Clark, C., F. Wilson and J. Bradley (1969). 'Industrial location and economic potential in Western Europe', *Regional Studies*, III, 197–212.

Deakin, B. M. and T. Seward (1969). *Productivity in Transport. A Study of Employment, Capital, Output, Productivity and Technical Change*, Cambridge University Press, Department of Applied Economics, Occasional Papers No. 17.

Deakin, B. M. and T. Seward (1970). Private communication of data used for their 1969 study.

Edwards, S. L. (1967). *The West Cumberland Transport Survey*, Northern Economic Planning Board.
Edwards, S. L. (1969). 'Transport costs in the wholesale trades', *Journal of Transport Economics and Policy*, III, 3, 272–8.
Edwards, S. L. (1970a). 'Transport cost in British industry', *Journal of Transport Economics and Policy*, IV, 3, 1–19.
Edwards, S. L. (1970b). Private communication.
Harris, C. D. (1954). 'The market as a factor in the localization of industry in the United States', *Annals*, Association of American Geographers, 44, 4, 315–48; reprinted in R. H. T. Smith, E. J. Taaffe and L. J. King (eds.), *Readings in Economic Geography. The Location of Economic Activity*, 1968, Rand McNally.
Isard, W. (1956). *Location and Space-Economy. A General Theory Relating to Industrial Location, Market Areas, Trade and Urban Structure*, Wiley.
Lösch, A. (1954). *The Economics of Location*, Yale University Press (translated by W. H. Woglom).
Luttrell, W. F. (1962). *Factory Location and Industrial Movement: A Study of Recent Experience in Great Britain*, National Institute of Economic and Social Research, 2 vols.
McCrone, G. (1969). *Regional Policy in Britain*, Allen and Unwin.
Ministry of Transport (1964). *Survey of Road Goods Transport 1962 Final Results; Part 1* and *Commodity Analysis*, HMSO. (*Geographical Analysis* was published in 1966.)
Myrdal, G. (1957). *Economic Theory and Under-Developed Regions*, Duckworth.
Needleman, L. and B. Scott (1964). 'Regional problems and location of industry policy in Britain', *Urban Studies*, I, 2, 153–73.
O'Sullivan, P. (1971). 'Forecasting interregional freight flows in Great Britain', in M. Chisholm, A. E. Frey and P. Haggett (eds.), *Regional Forecasting*, Butterworths.
Rimmer, P. J. (1970). *Freight Forwarding in Australia*, Department of Human Geography Publication HG/4, Australian National University.
Scottish Council (Development and Industry) (1962). *Report on the Scottish Economy 1960–1961*, Scottish Council.
Weber, A. (1929). *Theory of the Location of Industries*, University of Chicago Press (translated by C. J. Friedrich).

Index

Washington, DC

leisure & culture DUNDEE

Working in Partnership
with Dund... ...ncil